GLOBAL JUSTICE AND CLIMATE GOVERNANCE

This series publishes ground-breaking work on key topics in the area of global justice and human rights including democracy, gender, poverty, the environment, and just war. Books in the series are of broad interest to theorists working in politics, international relations, philosophy, and related disciplines.

Studies in Global Justice and Human Rights
Series Editor: Thom Brooks

Published titles
Retheorising Statelessness
Kelly Staples
(July 2012)

Health Inequalities and Global Injustice
Patti Tamara Lenard and Christine Straehle
(August 2012)

Rwanda and the Moral Obligation of Humanitarian Intervention
Joshua J. Kassner
(November 2012)

Institutions in Global Distributive Justice
András Miklós
(February 2013)

Human Rights from Community
Oche Onazi
(June 2013)

Immigration Justice
Peter W. Higgins
(August 2013)

The Morality of Peacekeeping
Daniel H. Levine
(December 2013)

International Development and Human Aid
Paulo Barcelos and Gabriele De Angelis
(September 2016)

Global Justice and Climate Governance
Alix Dietzel
(January 2019)

www.euppublishing.com/series/sgjhr

GLOBAL JUSTICE AND CLIMATE GOVERNANCE

BRIDGING THEORY AND PRACTICE

Alix Dietzel

EDINBURGH
University Press

This book is dedicated to my students – past, present, and future. You inspire me to write clearly, think deeply, and always stay curious.

Edinburgh University Press is one of the leading university presses in the UK. We publish academic books and journals in our selected subject areas across the humanities and social sciences, combining cutting-edge scholarship with high editorial and production values to produce academic works of lasting importance. For more information visit our website: edinburghuniversitypress.com

Edinburgh University Press Ltd
The Tun – Holyrood Road, 12(2f)
Jackson's Entry
Edinburgh EH8 8PJ

First published in hardback by Edinburgh University Press 2019

Typeset in 11/13 Palatino LT Std by
IDSUK (DataConnection) Ltd, and
printed and bound by CPI Group (UK) Ltd,
Croydon, CR0 4YY

A CIP record for this book is available from the British Library

ISBN 978 1 4744 3791 2 (hardback)
ISBN 978 1 4744 3792 9 (paperback)
ISBN 978 1 4744 3793 6 (webready PDF)
ISBN 978 1 4744 3794 3 (epub)

CONTENTS

Acknowledgements   vi
Abbreviations   viii
Introduction   1

Part I   Developing a Climate Justice Account

1   The Scope of Climate Justice   25

2   The Grounds of Climate Justice   41

3   The Demands of Climate Justice   58

Part II   Assessing Climate Governance

4   Bridging Theory and Practice   93

5   Assessing Multilateral Climate Governance   117

6   Assessing Transnational Climate Governance   161

Conclusion   202

References   220
Index   231

ACKNOWLEDGEMENTS

Writing this book would not have been possible without the ongoing support of my friends, family and colleagues. First, I would like to express my gratitude to the University of Sheffield Politics Department, where I spent the better part of a decade obtaining my undergraduate, master's and postgraduate degrees. My time in Sheffield no doubt helped shape my ideas, worldview and work ethic. Thank you in particular to Hayley Stevenson for the advice and support she provided when supervising my PhD. Her input has been invaluable in carrying my research forward. I would also like to thank Alasdair Cochrane for his brilliant insights during my viva and for his exemplary knack for finding typos (even in footnotes!). Finally, I am grateful for the friends I made during my time in Sheffield, whose insights helped shape my views on research, teaching and life as an academic. Special mention goes to Clara Sandelind-Stafford for reading the introduction of this book when it first started to take shape. Your support helped me to carry on writing.

Next, I would like to thank David Held, who examined my PhD and encouraged me to publish this book. David has been an inspiration since my undergraduate days when I first picked up his work and discovered cosmopolitanism. His continued support is deeply appreciated. I would also like to thank my fellow climate justice scholars for discussing ideas found in this book with me at various academic events – especially Simon Caney, Stephen Gardiner, Catriona McKinnon, Steve Vanderheiden, Rob Lawlor, Megan Blomfield and Henry Shue. In addition, I want to express my gratitude to the publishing team at Edinburgh University Press, the two anonymous reviewers of this book, and of course to Thom Brooks for inviting me to be part of the Studies in Global Justice and Human Rights series.

Finally, I would like to thank the School of Sociology, Politics and International Studies (SPAIS) at Bristol, who have taken me in and allowed me to flourish as a young scholar. I am grateful for the friends I have made here, especially Tim Fowler, who is always happy to discuss matters of justice, as well as Ashley Dodsworth, Hannah Parrott, Egle Cesnulyte and Rob Yates, who helped me survive my TA years.

Above all, I want to thank my family (the Rogan-Dietzel-Schmitz-Cletheroe collective) for their never-ending enthusiasm and encouragement. Even though you might not always understand what I am doing – you know that 'I am putting effort in the right direction,' as my brother Daniel Dietzel would say. Thank you in particular to my husband, Lewis Cletheroe. You have been unwavering in your encouragement, love and support for me. I am grateful for all that you do.

ABBREVIATIONS

100RC – 100 Resilient Cities
ACCCRN – Asian Cities Climate Change Resilience Network
ADP – Ad-Hoc Working Group on the Durban Platform for Enhanced Action
APA – Ad-Hoc Working Group on the Paris Agreement
APP – Asia-Pacific Partnership on Clean Development and Climate
ATP – Ability to Pay
AWG-LC – Ad-Hoc Working Group on Long-Term Cooperative Action
BRICS – Brazil, Russia, India, China, South Africa
CAN – Climate Action Network
CBDR – Common but Differentiated Responsibility
CSLF – Carbon Sequestration Leadership Forum
CDM – The Clean Development Mechanism
CO_2 – Carbon Dioxide
COP – Conference of the Parties
EU – European Union
ENB – Earth Negotiations Bulletin
GCF – Green Climate Fund
GHGs – Greenhouse Gases
GDP – Gross Domestic Product
ICHR – International Council on Human Rights
INDCs – Intended Nationally Determined Contributions
IPCC – Intergovernmental Panel on Climate Change
LEG – Least Developed Country Expert Group
LMDCs – Like Minded Group of Developing Countries
LPAA – Lima-Paris Action Agenda

MENA – Middle East and North Africa
NAZCA – Nonstate Actor Zone for Climate Action
NAPAs – National Adaptation Programmes for Action
NAPs – National Adaptation Plans
NGOs – Non-Governmental Organisations
OECD – Organisation for Economic Co-operation and Development
PATP – Polluter's Ability to Pay
PPP – Polluter Pays Principle
RGGI – Regional Greenhouse Gas Initiative
SSA – Sub-Saharan Africa
UDHR – Universal Declaration of Human Rights
UNDP – United Nations Development Programme
UNEP – United Nations Environment Programme
UNFCCC – United Nations Framework Convention on Climate Change
UNHCR – United Nations High Commissioner for Refugees
UNSG – United Nations Secretary General
VCS – Verified Carbon Standard
WHO – World Health Organization

INTRODUCTION

Climate change is one of the most significant challenges our global community has ever faced. As scientific evidence continues to accumulate, it is becoming obvious that climate change requires an urgent global response. Without such a response, rising sea levels, severe weather patterns and the spread of deadly diseases threaten the lives of both present and future generations. Although states continue to work together to act on climate change, most recently by ratifying the Paris Agreement, the global response has so far been woefully inadequate. Some world leaders, most notably United States president Donald Trump, seem outright determined to halt the climate change response altogether. With temperatures rising and climatic changes taking hold across the world, it is increasingly apparent that not enough is being done.

This book aims to make sense of the lack of proper response to climate change – focusing on what has gone wrong, what has gone right, and what might change now that the Paris Agreement has been ratified. Of course, one cannot assess what is going right and wrong without knowing what a right, or just, response to the climate change problem would look like in the first place. The book contends that cosmopolitan thinking on global justice is ideally placed for developing an understanding of what such a response requires. Cosmopolitans put individuals at the centre of moral concern and emphasise the importance of fair treatment of humans across the globe. Engaging in cosmopolitan global justice debates means exploring ethical aspects of the climate change problem: identifying victims of injustice, defining a fair distribution of climate responsibilities and assigning duties of justice to those who are responsible. In exploring these ethical aspects of the climate change problem, the book is able to determine what a just response to the climate change problem entails. Once this has been established, the book can then assess whether the current response to the climate change problem is, in fact, just.

Although this is not the first book on the subject of cosmopolitan global justice and climate change (or climate justice for short),[1] it is the first to normatively evaluate multilateral (state) and transnational (non-state) climate change responses, and in doing so make sense of the 'big picture' of climate change (mis)management and the injustices that come along with it. To date, cosmopolitan scholars have predominantly focused on multilateral rather than transnational responses to climate change. This is problematic, because transnational actors, including cities, corporations and non-governmental organisations, have become increasingly important in the past decade, mainly due to their significant potential to contribute to the climate change response. Their role is especially essential at this present moment because the Paris Agreement presents a significant shift away from a purely state-led approach. The post-Paris Agreement climate change regime does not see transnational actors as merely a 'helpful addition', but as a core element of action on climate change (Hale 2016: 14).

This raises important questions about the future of the climate change response, including ethical concerns around what a just response that encompasses transnational actors would look like. It is therefore vital to take stock of the very recent changes presented by the Paris Agreement. Cosmopolitan climate justice scholars should be particularly interested in how transnational climate change responses compare to multilateral climate change ones in terms of just distribution of burdens, allocation of responsibility, protection of humans and procedural justice. In addition, those who are concerned about climate politics more broadly should be interested to know what the increased role of non-state actors might mean for climate change action in the near and distant future. For example, has the climate change response become more efficient, or more chaotic, through the addition of non-state actors? Will non-state actors inspire states to do more, or will states become complacent in their reliance on non-state actors? These are questions the book seeks to answer by evaluating and comparing the multilateral and transnational climate change response from a cosmopolitan point of view.

While non-state actors are no doubt playing an increasingly significant role, the multilateral regime under the United Nations Framework Convention on Climate Change (UNFCCC) is by no means losing momentum. The Kyoto Protocol remains in force until 2020, at which time it is expected to be superseded by the recently ratified Paris Agreement. If anything, the multilateral regime is growing in importance, as more and more states commit to significant mitigation and adaptation targets. Assessing the multilateral response to climate change therefore remains paramount for climate justice scholars. Questions to ask of this response include whether the Paris Agreement presents a meaningful and significant step forward, whether allowing individual states to set their own targets will increase participation in mitigation and adaptation efforts, and whether the newly conceived five-year review process will ensure that these voluntary targets are met. These questions are especially important while the particulars of implementing the Paris Agreement are being negotiated – there is still time for change. In this sense, it remains crucial to investigate multilateral responses alongside transnational ones.

There has so far been no attempt to comprehensively assess and compare multilateral and transnational climate change governance responses from a cosmopolitan climate justice perspective. This is because cosmopolitans most often focus on multilateral responses (Vanderheiden 2008; Harris 2010; Lawrence 2014), but also because the ethical dimensions of climate change governance remain under-explored by climate governance scholars, particularly among those who study transnational climate change governance (Hoffman 2011; Bulkeley et al. 2014). Although transnational climate governance scholars are beginning to turn their attention to climate justice (Bulkeley, Edwards and Fuller 2014; Hale 2016; McKendry 2016), there is a lack of systematic research that normatively evaluates transnational climate change governance processes or explores how these processes compare to multilateral efforts. In other words, there remains a significant gap between theory and practice.

The book aims to bridge this gap by illustrating that climate justice theory can be used to assess climate change governance practice. Part I of the book develops a climate justice account that can be used to evaluate the global response to climate change. Because there has so far been no attempt to normatively assess and compare multilateral and transnational climate change governance, the book must provide a detailed account of climate justice to make this assessment possible in the first place. As was alluded to above, one cannot assess what is going right and wrong if one does not define what the right, or rather just, response to climate change would look like in the first place.

The first three chapters of the book therefore focus on answering four normative questions: who will be most affected, what exactly is at stake, what action must be taken in the face of climate change, and who should be responsible for this action. In answering these questions, the book sets out the scope, grounds and demands of climate justice in Chapters 1, 2 and 3 respectively. In this sense, Part I of the book pinpoints what exactly is normatively expected of the climate change response. As will be explained, climate change governance actors must: 1) protect the human right to health; 2) allocate state responsibilities according to wealth and emissions levels; and 3) ensure all capable non-state actors are held to account. It is only once these three demands have been defined that the book can turn to evaluating the climate change response in Part II.

Part II assesses how multilateral and transnational climate change actors have fared in meeting the demands set out in Part I. Chapter 4 explains how the global response will be assessed. The book grounds the responsibility of multilateral and transnational climate change governance actors in their ability to restructure the social and political context to enable the three demands of justice defined in Part I and sets out a four-point hierarchy that can be used to investigate to what extent global governance actors enable each demand. Chapters 5 and 6 then turn to the assessment of multilateral and transnational climate change responses, respectively. These chapters will pinpoint what has gone right, what has gone wrong and what has changed since the Paris Agreement was ratified. Chapter 5 focuses on the key policy decisions made by multilateral actors, including the Convention, Kyoto Protocol and Paris Agreement, as well as investigating how these decisions were negotiated. Chapter 6 takes a slightly different approach, focusing on ten key examples of non-state climate change responses to get a sense of how transnational actors are addressing the climate change problem.

Although some of the findings made in Chapters 5 and 6 are intuitive, because it is well known that climate change governance actors are not adequately addressing the climate change problem, it is nevertheless valuable to systematically evaluate the hindrances facing these state and non-state actors. This allows for research to go beyond intuition and assumption and provide specialised and detailed knowledge of the global response to climate change. The importance of this knowledge cannot be understated, because intuitive thinking may not be adequate for explicating normative suggestions for reform toward a better condition of justice.

A thorough examination of current practices provides a common denominator from which to begin suggesting what is needed to ensure a just response to climate change. The Conclusion of the book will therefore focus on the lessons that can be learned from a climate-justice-focused assessment and comparison of multilateral and transnational climate governance. It is here that the book puts forward policy recommendations for the post-Paris Agreement era.

By providing a cosmopolitan assessment of the multilateral and transnational climate change response, the book aims to bridge the gap between climate justice and climate governance scholars. In doing so, the book provides room for further discussion between these two fields, which have been traditionally concerned with complementary yet separate research agendas. The book demonstrates that bridging the fields of climate justice and climate governance can underwrite future research and ultimately help to bring about a more just global response to climate change. More broadly, the book hopes to inform readers about the recent changes in climate change politics and what these might mean in terms of a fair and effective response to the climate change problem now that the Paris Agreement has been ratified. This will allow for an in-depth normative understanding of the climate change response. With these wider aims in mind, this first introductory chapter begins with a brief overview of the climate change problem. Next, the chapter will defend the use of a cosmopolitan approach and comment on existing cosmopolitan research, to explain how the book relates to and contributes to current work on the subject. Finally, the chapter will outline what is to come in the remainder of the book.

THE CLIMATE CHANGE PROBLEM

Climate change has been under investigation by scientists for decades. Each year, these experts grow more certain of their predictions and present increasingly nuanced and detailed evidence to back up their claims. The complexity of the climate change problem makes it difficult to summarise in such a short space. For this reason, the discussion below will focus on three key aspects of the problem: causes, consequences and action required. This focus allows the book to present key issues for further exploration: who will be most affected, what exactly is at stake, what action must be taken in the face of climate change and who should be responsible for this action. These are all issues of justice, as will be explained below.

Causes of Climate Change

It is no longer a question that humans are the predominant cause of climate change. The latest report by the IPCC (2014a: 5) expressed this in immutable terms: there is a 95% chance that human influence has been the dominant cause of climate change since the mid-twentieth century, a fact that 97% of scientists around the world agree on. This is a monumental level of certainty considering that it is based on the research of thousands of scientists working independently from one another – the latest IPCC report involved 800 experts and was based on the analysis of over 9,000 peer reviewed papers. The manner in which humans cause climate change is also no longer up for debate. Scientists agree that humans cause climate change by increasing naturally occurring greenhouse gases (GHGs).[2] GHGs are extremely long-living in nature. This means that present-day emissions will continue to exist in the atmosphere for centuries to come (IPCC 1990: 52). As they continue to accumulate, GHGs lead to warming of the atmosphere and surface of the earth. Eventually, the atmospheric concentrations of GHGs become so high that they lead to irreversible changes in the climate or, as it is more commonly referred to, climate change (IPCC 1990: 53).

Consequences of Climate Change

The consequences of climate change raise profound questions about justice. For example, consider the fact that climate change will primarily affect two groups of victims: future generations, and those living in less developed countries. Future generations will bear the brunt of climate change effects because the most severe consequences of clime change will not occur until the 2°C temperature threshold is crossed sometime between 2050 and 2100 (IPCC 2014a: 14), implying that many of the worst effects will hit those who are yet to be born. In this sense, the primary victims of climate change did not cause the problem. This presents issues of justice – is it just that those who did not cause the problem must suffer the consequences? What might be owed to future generations, in terms of protection from climate change or compensation for having to live with the consequences?

Although future generations will bear most of the burden of climate change, this burden will not be evenly distributed across the globe. In fact, it is estimated that less developed countries will bear 75–80% of the burden of climate change (World Bank 2010: xx). This is for two broad reasons. First, evidence increasingly points to the fact that areas of the world less developed countries call home face greater risk (IPCC 2007: 65). For example, agricultural production, including access to food, is projected to be severely compromised in many African countries (IPCC 2007: 50) and sea level rises are projected particularly for Asian mega deltas and for small island communities (IPCC 2007: 65). This is in contrast with richer parts of the world. In New Zealand, initial benefits of climate change, such as longer growing seasons, are projected in some regions (IPCC 2007: 50). Similarly, in North America, climate change is projected to increase aggregate yields of rain-fed agriculture by 5–20% in the first few decades of warming (IPCC 2007: 50).

Of course, richer regions will also eventually be adversely affected by climate change, but not as severely as poorer ones, because these regions have lower adaptive capacity. This is the second reason less developed countries will be harder hit by climate change. Less developed countries often have other problems, such as poverty or weak infrastructure, which create conditions of low adaptive capacity to climate change (IPCC 2001: 12). Less developed countries may not have the financial capacity or infrastructure necessary to combat the ill effects of climate change, and will, as a consequence, not be able to prepare or defend themselves against effects like flooding, droughts or rising sea levels as effectively as richer states. Again, this raises profound questions of justice – is it fair that less developed countries must suffer from climate change if they have not caused it, and should richer countries be held responsible for adaptation costs in these countries?

Further justice concerns emerge when considering the effects of climate change in more detail. The first, and perhaps most well-known, consequence of climate change is warming of the planet, which is an ongoing effect carefully observed by scientists over the last few decades. According to the latest IPCC report, warming of the climate system is now considered unequivocal (IPCC 2014a: 3). Warming has been detected in changes in surface and atmospheric temperatures as well as in temperatures of the upper hundred metres of global oceans (IPCC 2007: 39). The most up-to-date findings suggest that these observed changes in temperature are unprecedented: each of the last three decades has been successively warmer at the earth's surface than any preceding decade since 1850 (IPCC 2014a: 3). The IPCC does not project that this warming will slow down in the coming decades and centuries, and estimates that if current emissions trends continue, global temperatures will pass the threshold of 2°C warming above pre-industrial levels sometime between 2050 and 2100. It is at this point that climate change becomes 'dangerous' (IPCC 2014a: 14).

Although temperatures have not yet reached 2°C, the 1°C threshold was surpassed in 2016, and this has had observable effects on weather patterns. One such observable effect is the widespread melting of ice. According to the IPCC (2014a: 5), over the last two decades, the Greenland and Antarctic ice sheets have been losing mass, glaciers have continued to shrink almost worldwide, and Arctic sea ice and Northern Hemisphere spring snow cover have continued to decrease. The widespread melting of ice is causing sea level rises. From 1901–2010, global mean sea levels rose by 0.19 metres (IPCC 2014a: 6). This rise is significant, as the rate of sea level rise since the mid-nineteenth century has been larger than the mean rate during the previous two millennia (IPCC 2014a: 6). Global mean sea levels are predicted to continue to rise during the twenty-first century and could rise by a further 0.98 metres by 2100 (IPCC 2014a: 18).

If the temperatures continue to rise, ice continues to melt and sea levels continue to rise, there will be severe knock-on effects. These will include increased instances of heavy precipitation, floods, droughts and heatwaves (IPCC 2007: 53). Floods will damage crops, cause soil erosion, result in the inability to cultivate land due to waterlogging of soils, have adverse effects on quality of surface and groundwater, lead to the contamination of the water supply and present an increased risk of death, injuries and infectious, respiratory and skin diseases (IPCC 2007: 48). Increased droughts, on the other hand, will result in land degradation, lower yields of crops due to crop damage and failure, increased livestock deaths, increased risks of wildfire, more widespread water stress, higher risk of food and water shortage, increased risk of malnutrition and amplified risk of water and food-borne diseases (IPCC 2007: 48). Heatwaves will also have destructive consequences, including reduced agriculture yields in warmer regions due to heat stress, increased danger of wildfires, higher water demand, water quality problems, increased risk of heat-related mortality, and poor quality of life for those without appropriate housing (IPCC 2007: 48).

A long list of consequences sometimes makes it difficult to picture what exactly this will mean for human beings. It is also difficult to pin down clear priorities for climate justice. For example, should health be prioritised over shelter? Is clean water more important than nutritious food? How should responsibility for these effects be distributed? These questions of justice, along with the ones briefly outlined above, will all be addressed in the first part of the book, in Chapters 1, 2 and 3. Chapter 1 focuses on how to include those most affected by climate change (future generations, and those living in less developed countries) in the realm of moral concern. Chapter 2 discusses what exactly is at stake for these victims (the human right to health), and Chapter 3 then turns to the question of what action must be taken in the face of climate change, and who should be responsible for this action, a topic that is briefly introduced below.

What Action is Required?

The IPCC (2014a: 14) estimates that because economic growth is set to persist, and global population is set to increase, mean global surface temperatures could rise by as much as 4.8°C by 2100 without additional efforts to reduce GHG emissions beyond those in place today. This level of change in temperature would have severe and irreversible consequences, including mass extinction of animals, changes in marine ecosystem productivity, damage to fisheries, changes in oceanic oxygen concentrations and decreased terrestrial vegetation (IPCC: 2007: 53). Any of these changes would have extreme detrimental impacts on the human population (IPCC: 2007: 53). It is therefore paramount that action is taken sooner rather than later. What action to take against climate change will be discussed in detail in Chapter 3, which will: 1) define an emissions threshold required to protect human health; 2) outline which states are most responsible for lowering emissions and/or contributing to climate costs; and 3) identify non-state actors who are responsible for action on climate change. For now, this introductory chapter outlines a few general assumptions that can be made about what kind of action will be required.

The first assumption is that both adaptation and mitigation will be necessary to combat climate change. Mitigation refers to cutting back on emissions, and adaptation refers to tactics that can be taken to adjust to climate change effects. Examples of mitigation include use of renewable energy, changes in consumption patterns, increased fuel efficiency or using biofuels, making new buildings more sustainable and making use of waste as fuel. With these types of changes, it is possible to significantly lower emissions, reducing the risks of climate change (IPCC 2007: 65). The IPCC maintains, at the time of writing, that there are multiple mitigation pathways that are likely to limit warming to below 2°C. These pathways would require substantial emissions reductions over the next few decades: emissions would have to be cut by 40% and 70% by 2050 compared to 2010, and would need to be near zero or below in 2100 (IPCC 2014a: 14).

It is important to note that there is high confidence that mitigation alone cannot avoid all climate change impacts (IPCC 2007: 65). Even if all emissions were halted, the emissions that already exist in the atmosphere would still cause changes in the climate. Therefore, adaptation measures will be required to cope with the effects of climate change. Examples of adaptation include water reuse, rainwater harvesting, adjustment of planting dates and crop variety, crop relocation, erosion control, building seawalls and storm surge barriers, creating heatwave action plans, protecting water supplies, inoculating populations against certain diseases and strengthening infrastructure. However, adaptation alone, like mitigation alone, is not expected to cope with all the projected effects of climate change, especially not over the long term as impacts increase in magnitude (IPCC 2007: 56). It is simply not possible to continue to emit at current levels and rely on adaptation alone – eventually irreversible damages such as the ones outlined above will make it impossible to 'adapt our way out'. For this reason, both adaptation and mitigation are necessary, and more importantly can work together to significantly reduce the risks of climate change (IPCC 2007: 65).

The second assumption that can be made about climate change action is that the response to climate change will necessarily have to be collective. Since its first report, the IPCC has insisted that climate change action will require a high degree of international cooperation (1990: 60). Climate change is a global problem that cannot be solved by a group of states or a group of willing individuals alone. The third and final assumption that can be made is that collective action will not only be required in the short term. The latest IPCC report claims that aspects of climate change will persist for many centuries even if emissions are stopped, which implies a substantial multi-century climate change commitment (IPCC 2014a: 19). Sustained global collective action will not be easy to achieve or maintain. However, it is important that action is taken sooner rather than later, in order to avoid irreversible damages. Part II of the book will explore what action has been taken so far, why this can be considered inadequate from a cosmopolitan perspective and what must be done in the future to meet the demands of climate justice. This will include a discussion on which actors are responsible for how much mitigation and adaptation. For now, the chapter turns to discussing the problem of scientific uncertainty.

A Note on Scientific Uncertainty

Scientific uncertainty is an unavoidable subject because climate change science is, at the time of writing, not considered to be indisputable. Each of the five reports produced by the IPCC discusses some degree of uncertainty. Predicting the future response of the atmosphere and climate systems is by no means a straightforward task, which is why a degree of uncertainty about the future remains. This degree of scientific uncertainty has been, in the past, used as a reason for inaction. Governments, most notably in the United States of America (USA), have refused to take part in global climate change deals, citing scientific uncertainty. This phenomenon extends beyond governments, as there are individuals who simply do not believe that there is enough evidence to be certain about the effects of climate change and therefore consider the urgency for action to be either exaggerated or outright untrue. A response to these critics of climate science would be to assert that the book is based on best available evidence, provided by respected scientists who are experts in their fields. Nevertheless, it is important to meet potential critics head on because misinformation on the topic of uncertainty should not be a cause for inaction. The climate change problem is simply too important for continued inaction to be morally acceptable.

As was explained above, 97% of climate scientists are certain that climate change is predominantly caused by humans (IPCC 2014a: 5). The problem of uncertainty therefore no longer lies with the question of whether humans cause climate change, but rather how severe the consequences of climate change will be. Even though there are undoubtedly uncertainties involved in the modelling of future scenarios, it is important to stress that these uncertainties are not as widespread or significant as sceptics make them out to be. For one, there is a level of certainty that arises from the fact that climate change effects are already occurring. The IPCC (2007: 39) is 'virtually certain' (99–100%) that surface and atmospheric temperatures, as well as the temperatures of the upper hundred metres of global oceans, have warmed over the last decades. Furthermore, the IPCC (2014a: 3) has 'high confidence', meaning that there is 'robust evidence and high agreement', that this warming has resulted in a diminishing of ice and snow, and a rise in sea levels. The fact that these changes are already observable gives weight to the level of certainty that GHG emissions have, and will continue to have, an effect on the global climate.

In addition, the IPCC reports contain a high level of certainty about many of the most important predicted effects. For example, it is 'very likely' (90–100%) that heatwaves will occur more often and last longer, and that extreme precipitation events will become more intense and frequent in many regions (IPCC 2014a: 8). The IPCC (2014a: 9) is also 'virtually certain' (99–100%) that near-surface permafrost extent at high northern latitudes will be reduced as global mean surface temperature increases. In addition, the IPCC (2014a: 9) predicts that global mean sea-level rise will continue during the twenty-first century, 'very likely' (90–100%) at a faster rate than observed from 1971 to 2010. The level of certainty around these effects is very high, which is significant when considering how many thousands of eminent scientists have been involved in reaching a consensus on the likelihood of these effects.

In sum, the key effects of climate change are already occurring, and thousands of scientific experts are very much certain that they will continue to worsen. Even though there is some uncertainty about how serious the effects will be, it is undisputed that climate change will have detrimental effects. For this reason, it is no longer undisputed that humans will suffer – uncertainty merely lies in the question of how many will suffer. This book takes the cosmopolitan view that human suffering is morally important, and that it is therefore imperative to act on climate change (Shue 2010: 147). On this cosmopolitan view, waiting for absolute certainty on the amount of human suffering is not required for action, because certainty that humans will suffer already exists. The chapter now turns to explaining why such a view has been taken in the first place.

WHY COSMOPOLITANISM?

Cosmopolitan scholars assume that all humans have equal moral worth and the right to equal moral consideration (Dietzel 2017: 91). This focus on the moral importance of the individual has led some cosmopolitans to critically engage with theories of justice, which were traditionally confined to the state. These endeavours led to emergence of the discipline of global justice, in which scholars focus on what individuals across the world deserve and how distribution of these entitlements can be achieved (Dietzel 2017: 91). The book takes such a cosmopolitan global justice approach for two key reasons. First, climate change is a problem of global justice by its very nature. This makes the cosmopolitan approach particularly useful because it presents a problem that global justice scholars can readily engage with. Second, the cosmopolitan focus on individuals and fairness are excellent tools for analysis in the case of climate change, which affects individuals across the globe and results in extreme levels of unfairness in terms of burden sharing. Global justice is very well suited for thinking through such problems.

Let us first consider why climate change can be considered a global justice problem. First, it is undoubtedly a truly global problem. Emissions cannot be confined within states, instead rising into the atmosphere and causing global temperature changes within and outside of their original state borders. If an individual drives a car in France, their emissions will contribute to the overall level of emissions in the atmosphere, which will cumulatively cause problems across the world. For this reason, emissions from one individual in France could affect another individual in Indonesia whose house is flooded as a result of high emissions levels. Although it is difficult to determine direct blame or fault, it is nonetheless undeniable that all individuals, or states, or corporations who emit GHGs contribute to climate change. In this sense, the global nature of the climate change problem defies conventional assumptions about state sovereignty and geographically bounded justice (Vanderheiden 2008: xiv), which makes it a truly global problem that raises global ethical concerns.

Secondly, the world's nations and peoples, both present and future, depend on a global scheme of cooperation to respond to climate change, and are therefore part of a global justice community (Vanderheiden 2008: 105). No one nation or individual can solve climate change on their own. The European Union (EU), for example, only accounts for around 10% of global emissions, and the USA accounts for around 16%. Even if all countries in the EU and all states in the USA stopped emitting overnight, this would not be enough to halt climate change if no other country reduced their emissions, since global emissions must be lowered by 40–70% by 2050 and be at or near zero by 2100. There is no doubt that combating climate change will require a collaborative effort, implying the need for global agreements. Coming to such agreements will inevitably involve discussion about which actors must lower emissions by how much or which actors should contribute to the costs of climate change. These are, by their very nature, questions of distributive justice. Thus, while critics of global justice may be right to hold that global interdependence does not arise among humans in all instances, they would surely be wrong to deny that humans are interdependent in their common reliance on combating climate change (Vanderheiden 2008: 105).

Third, climate change presents an unfair distribution of benefits and burdens on a global level. Climate change will most negatively affect those living in less developed countries who have done the least to contribute to the causes of climate change, while those living in developed countries, who have contributed the most emissions, will suffer the least. As discussed above, this is because less developed countries are located in areas which will bear most of the problems associated with climate change, and furthermore because these countries do not always have the ability to adapt to dangerous weather patterns. Developed countries, on the other hand, are located in areas which are not predicted to suffer extreme weather conditions to the same extent or as quickly. Furthermore, developed countries often have the capability of coping with weather changes because of their existing institutional and financial power. In this sense, climate change clearly presents a case of global injustice, and the empirical conditions of climate change 'cry out for justice' (Harris 2010: 37), as climate change is imposed on people who are already poor, cannot adequately protect themselves and have no real say in the matter. Climate change, then, fits into the remit of global justice quite naturally. Scholars from the discipline are therefore very much equipped to tackle the ethical complexities of the climate change problem.

On top of this, the cosmopolitan focus on individuals and fairness is highly useful for the assessment of the climate change problem. As Harris (2013: 129) has pointed out, a global justice approach 'can help us to frame, visualise, and talk about climate politics, and to formulate and implement related policies'. The importance of the moral worth of the individual is particularly useful because climate change is rife with ethical issues that concern individuals. GHG emissions are not confined to states but can be caused by any individual, regardless of their place of birth. Similarly, climate change can affect any individual in any state. This implies that individuals are both the victims and cause of climate change, raising distributive questions about which individuals should be protected and which individuals may have to refrain from emitting. The cosmopolitan position is primed to addressing these questions head on.

Furthermore, as discussed above, the primary victims of climate change will be future generations and those living in less developed countries. An understanding of the equal moral worth of humans allows global justice scholars to argue that we need greater collective action on climate change in order to rectify the unequal treatment of morally equal human beings. This understanding of, and focus on, fair burden sharing allows for an exploration of what equitable distribution might look like in the case of climate change. Finally, the concern for what is fair and just implies being critical of the status quo. Cosmopolitan global justice scholars are interested in reforming institutions, laws and procedures (Gardiner 2010: 64). This inevitably involves changing structures and systems, which is very important in the case of climate change, because combating the problem will require substantial changes in behaviour at every level – from the individual to the state. For all of the reasons above, the book will take a cosmopolitan climate justice approach, contributing to a discipline that has developed very rapidly in the past decade. This development is briefly reflected on below.

Existing Cosmopolitan Climate Justice Research

In 2004, Stephen Gardiner observed that 'very few moral philosophers have written on climate change' (Gardiner 2004a: 555). Almost fifteen years later, this is undoubtedly no longer the case. In fact, literature on the ethics of climate change has flourished since Gardiner's seminal piece. Cosmopolitan climate justice is one strand of climate ethics that has undergone particularly interesting developments. Initially, scholars confined themselves to abstract normative considerations. Much of the early work was focused on assessing principles for how the costs of mitigation, adaptation and compensation should be distributed between wealthier and poorer regions and between present and future generations.[3] This initial theorising was highly important in terms of building a foundation others could carry forward for the assessment of current practice. Such assessment has recently become much more popular among global justice scholars, perhaps as a result of continued inaction on climate change. In a sense, the ongoing failure to adequately respond to climate change has made it increasingly clear that climate justice scholars 'must pay much more attention to the questions of how to actually put effective and equitable climate governance into practice' (Maltais and McKinnon 2015: x).

One of the earliest examples of an attempt to assess current practice and make recommendations for change can be found in Paul Harris's 2010 *World Ethics and Climate Change*. This book focuses on individual responsibility and aims to evaluate the multilateral climate change regime from a cosmopolitan perspective. Harris's 2013 follow-on book, *What's Wrong with Climate Politics and How to Fix It*, puts forward concrete suggestions for the multilateral regime. However, Harris does not attempt to set out his own cosmopolitan position in either book, instead drawing on existing accounts and using these to make recommendations for change. By contrast, Peter Lawrence's 2014 *Justice for Future Generations* develops a climate justice account based on a discussion of what future generations deserve. This account is then applied to assess the multilateral response to the climate change problem, particularly the UNFCCC and international law more broadly. There are no other monographs on the subject, and none that assess transnational governance, but recently there has been somewhat of an explosion of shorter pieces and edited volumes that assess the multilateral regime from a climate justice perspective.

One of the most recent edited volumes is Clare Heyward and Dominic Roser's 2016 *Climate Justice in a Non-Ideal World*, which centres around engaging with the 'practical context' and attempting 'to arrive at action-guidance' (Heyward and Roser 2016: 1). Another key example of an edited volume that focuses on assessing current practice is Jeremy Moss's 2015 *Climate Change and Justice*, which focuses on 'assessing the dangers we face' from a climate justice perspective in 'new and interesting ways' (Moss 2015: 1). The third and final example of an edited volume is Aaron Maltais and Catriona McKinnon's 2015 *The Ethics of Climate Governance*, which focuses on assessing four key areas of climate governance: vulnerability and domination, tensions between democratic values and effectiveness, the challenges of motivating present generations, and new technologies (Maltais and McKinnon 2015: x).

Alongside these edited volumes, there has also been a proliferation of papers exploring the multilateral regime from a climate justice perspective. For example, Steven Vanderheiden (2015) has taken on the Green Climate Fund; Chukwumerije Okereke and Philip Coventry (2016) have provided an account on the evolution of the UNFCCC; and Marcelo Santos (2017) has offered an assessment of the Paris Agreement. It is clear that there is an emerging trend in the literature to assess current practice. However, although cosmopolitans have begun to explore the 'real world' of climate change politics, most scholars focus on multilateral rather than transnational responses to climate change. This is problematic when considering that the post-Paris Agreement regime seems to place transnational actors at the heart of the climate change response (Hale 2016: 12). Questions on what this might mean in terms of a just response to climate change therefore largely remain unanswered.

Encouragingly, a handful of climate justice and climate governance scholars are beginning to explore transnational climate change governance. For example, Corina McKendry (2016) has recently written a paper that focuses how just city action on climate change in Chicago, Birmingham and Vancouver has been. Similarly, Harriet Bulkeley, Gareth Edwards and Sara Fuller (2014) have applied existing cosmopolitan climate justice scholarship in order to assess city-based responses to climate change. As a final example, Thomas Hale (2016) has considered what the Paris Agreement means for transnational climate change actors. Here, Hale begins to ask questions relating to climate justice. He wonders whether 'the pragmatic, problem-solving approach that brings business and sub-national jurisdictions into the regime risk[s] detracting from questions of climate justice and an equitable distribution of the atmosphere's resources' (Hale 2016: 20). In addition, Hale asks whether the normative questions that have fascinated climate justice scholars could apply to transnational climate change governance: for example, 'what is the historical responsibility of a city, and what is an equitable contribution from a for-profit corporation?' (Hale 2016: 20). Hale leaves these questions unanswered, setting out a post-Paris Agreement agenda rather than definitively creating a climate justice account on transnational governance.

This book aims to answer the questions Hale sets out, thereby adding to the growing body of literature concerning itself with the assessment of multilateral and transnational climate change governance. As was explained at the beginning of this Introduction and outlined above, there is a distinct lack of systematic cosmopolitan research that evaluates transnational climate change governance processes or explores how these processes compare to multilateral efforts. In this sense, there remains a significant gap between theory and practice. The remainder of the book aims to bridge this gap, and how exactly this will be achieved is outlined below.

STRUCTURE OF THE BOOK

This book assesses the global response to climate change, both multilateral (state) and transnational (non-state), from a cosmopolitan global justice perspective. In doing so, the book aims to bridge the gap between theory and practice by illustrating how the theory of climate justice can be applied to evaluate the practice of climate change governance. The book demonstrates that bridging this gap allows for a comprehensive normative understanding of the global climate change response that can underwrite future research and ultimately help to bring about a more just global response to climate change. To achieve the aim of bridging the gap between theory and practice, the book is divided into two distinct halves. The first half of the book sets out a climate justice account that has been specifically developed for the assessment of climate change governance, and the second half of the book evaluates multilateral and transnational climate change responses. These two halves unfold as follows.

Part I, 'Developing a Climate Justice Account', is divided into three chapters. Chapter 1 defines the scope of climate justice. Defining the scope of justice is an essential part of any global justice account, because doing so clarifies who must be included in considerations of justice, or in other words, how wide the net of justice should be cast. The chapter considers the merits of a non-relational versus relational approach to climate justice and argues that both relational and non-relational elements are necessary to include the two key groups of victims of climate change in the scope of justice: future generations and those living in less developed countries. In this way, the chapter defends, and begins to develop, a climate justice account that includes both relational and non-relational elements. This account is further developed in Chapter 2, which defines the non-relational grounds of climate justice, and Chapter 3, which applies the mixed position in order to develop three demands of climate justice.

Chapter 2 defines the grounds of climate justice. Defining the grounds of justice is a key task for any climate justice account because it allows readers to understand what must be normatively prioritised. The grounds of justice in this sense represent the moral underpinnings of the climate justice account, a normative subfloor that must not be crossed. The chapter makes the case for using the human right to health as the non-relational moral minimum that grounds the climate justice position. Chapter 2 puts forward that the human right to health provides a strong foundation for climate justice because it captures the threats climate change poses to humans more comprehensively than other key human rights, including the right to food and water, the right to life and the right to free movement.

Chapter 3 completes the climate justice account by defining three demands of justice required to meet a condition of justice in the case of climate change. These three demands are considered normative principles that must underwrite a more just global response to climate change. The chapter is organised into three parts, each one developing a demand of justice. The first part of the chapter concerns the right to health and sets out a minimum set of actions that must be pursued in order to protect this right. Part two of the chapter conducts relational analysis by exploring the relationship between developed countries and less developed countries, and puts forward that states should be held to account for climate change action according to both their emissions levels and wealth levels. Finally, the third part of the chapter conducts relational analysis of the relationship between those who cause climate change and those who suffer from its effects, and makes the case that responsible actors extend beyond states to all capable actors, including individuals, firms, sub-state entities, international institutions and states, irrespective of the country they live or exist in. Ultimately, the chapter defines the following three demands of climate justice:

The Three Demands of Climate Justice

1. a) Global temperature changes must be kept at or below 2°C.
   b) Adaptation must be prioritised alongside mitigation.
2. The distribution of benefits and burdens in global climate change action must be based on the Polluter's Ability to Pay (PATP) model.
3. Capable actors, including individuals, firms, sub-state entities and international institutions, irrespective of the country in which they live or exist, must be held responsible for lowering emissions and/ or contributing to adaptation efforts, in line with their respective capabilities.

It is only once these three demands have been defined that the book can turn to evaluating the climate change response in Part II, 'Assessing Climate Governance'. The assessment in Part II focuses on both multilateral and transnational climate governance and aims to pinpoint what has gone right, what has gone wrong and what has changed since the Paris Agreement was ratified. The overall aim is to provide a comprehensive overview and assessment of climate change governance.

Part II is organised into three chapters. Chapter 4 sets out the parameters for assessment. The chapter first provides an overview of the processes involved in global climate change governance: multilateral (United Nations Framework Convention on Climate Change, or UNFCCC) and transnational (cities, corporations, NGOs, sub-state authorities) climate change governance. Following this, Chapter 4 outlines why actors in the UNFCCC and actors involved in transnational governance processes can be held responsible for bringing about a just response to the climate change problem. The chapter grounds the responsibility of these actors in their capability to enable the three demands of justice set out in Chapter 3 by restructuring the social and political context. Finally, Chapter 4 outlines a methodological framework to clarify how current practice will be assessed. This framework is based on a four-point hierarchy (outlined below) that can be used to investigate to what extent global governance actors enable each demand of justice.

The Four-Point Hierarchy

1. Actors in the institution enable the demand of justice – the demand of justice is unequivocally fulfilled in its entirety.
2. Actors in the institution are consistently working towards enabling the demand of justice – the demand of justice is not yet fulfilled, but there are policies in place which are consistently leading towards this goal.
3. Actors in the institution have promised to begin working on enabling the demand of justice in the future – no policy has been adopted, but there is the potential for the creation of policy in order to consistently work towards enabling the demand of justice.
4. Actors in the institution do not enable the demand of justice – there has been no promise or attempt to enable the demand of justice and there are no policies in place.

Chapter 5 assesses to what extent multilateral actors enable the three demands of justice developed in Part I of the book by using the four-point hierarchy above. Taking each demand of justice in turn, the chapter focuses on normative commitments made in the Convention as well as assessing the policies set out in the Kyoto Protocol and examining what has been achieved so far by multilateral actors. Finally, the chapter assesses to what extent the Paris Agreement presents a shift from existing policies. In this way, the chapter provides a historical overview of multilateral climate change action, as well as looking to the future. Chapter 5 puts forward that although there has been some progress made, none of the demands of justice come close to being met, and that there is an urgent need for change in the multilateral regime.

Mirroring Chapter 5, Chapter 6 assesses to what extent transnational actors enable the three demands of climate justice by drawing on the four-point hierarchy above. The assessment makes use of both existing climate change governance research and ten examples of transnational climate change governance initiatives, providing an insight into how transnational climate change governance has developed and where it stands today. Chapter 6 focuses on one demand of climate justice at a time, assessing both what has been promised by transnational actors and what has been achieved so far. The chapter puts forward that although there is room for cautious optimism, overall transnational actors fail to fully enable any of the three demands of justice. The final part of Chapter 6 summarises the findings made in Part II and considers what role both multilateral and transnational actors might play in the post-Paris Agreement regime. This is expanded on in the Conclusion of the book.

The Conclusion focuses on the lessons that can be learned from the bridging of theory and practice. More specifically, the Conclusion considers how transnational and multilateral responses compare, and explains that state and non-state actors face very similar problems, including the ongoing struggle to lower GHG emissions at the rate required, the entrenched favouring of mitigation over adaptation, the pervasive exclusion of less developed countries from decision-making processes and the incessant failure to change the behaviour of responsible actors. These shared problems imply that integrating transnational climate change actors in multilateral processes, which the post-Paris Agreement regime is moving towards, may not be a simple or straight-forward improvement of the climate change response. The Conclusion therefore reflects on whether the direction the post-Paris regime is heading in might be a hindrance to a just response to the climate change problem, rather than a help. The book ultimately recommends that transnational actors should be given as much space as possible to pursue their ambitions, with limited guidance from the UNFCCC.

Although this book specifically aims to contribute to climate justice and climate governance literature by bridging the gap between theory and practice, the findings made here should also be of interest to those from outside of these academic disciplines. The current response to climate change is undoubtedly inadequate. It is important to explore why this is, and consider what can be changed, especially in the run-up to the Paris Agreement being implemented. The lives of millions, or potentially billions (depending on our actions), are at stake. Readers of this book should feel angered by the slow pace of change, yet encouraged that there is continuous movement forward. Most of all, it is my hope that readers are motivated to learn more about what they can do to support the global response to climate change, whether at the individual level, or by becoming climate advocates and pushing for change at all levels of the political spectrum. May this book arm you with information and inspire you to do more.

NOTES

1. See for example Vanderheiden 2008, Harris 2010, Lawrence 2014, Heyward and Roser (eds) 2016, Moss (eds) 2015, Maltais and McKinnon (eds) 2015.
2. Carbon dioxide, methane, chlorofluorocarbons and nitrous oxide are referred to under the umbrella term of GHGs.
3. See for example Caney (2005, 2009, 2010), Vanderheiden (2008), Shue (2014).

Part I

Developing a Climate Justice Account

Chapter 1

THE SCOPE OF CLIMATE JUSTICE

INTRODUCTION

The first part of this book focuses on developing a climate justice account that can be used for the evaluation of the global response to climate change. There has so far been no attempt to normatively assess and compare multilateral and transnational climate change responses. The book must therefore provide an account of climate justice to make this assessment possible in the first place. It is not feasible to simply apply or update an existing account, because there is no account that has attempted the kind of analysis conducted here. In order to develop a climate justice account, the book will focus on answering four key questions raised by the climate change problem: who will be most affected, what exactly is at stake, what action must be taken in the face of climate change and who should be responsible for this action? Chapter 1 concerns the first question – who will be most affected – by setting out a scope of justice, which clarifies who must be included in considerations of justice, or in other words how wide the net of justice should be cast. In the case of this book, the net is both wide and deep, because the scope of climate justice is both non-relational and relational. This allows the book to prioritise those most affected by climate change: future generations and those living in less developed countries.

Chapter 2 turns to the second question – what exactly is at stake – by setting out the grounds of climate justice, which allows readers to understand what must be normatively prioritised. In the case of this book, it is the human right to health, which forms a normative subfloor that must not be crossed. The final two questions – what action must be taken in the face of climate change, and who should be responsible for this action – will be the subject of Chapter 3, which sets out three demands of climate justice. It is important to define what climate justice demands because this clarifies what exactly is normatively expected from the global response to climate change. In the case of this book, the demands centre around who should be responsible for climate change and what action must be taken. The first demand defines what must be done to protect the human right to health, the second puts forward that burdens must be distributed fairly between states and the third requires that all capable actors are held to account. By carefully exploring the scope, grounds and demands of climate justice, the book develops a new climate justice account specifically designed to be used for the evaluation of climate change governance. It is only once these fundamental elements of the climate justice account have been defined that the book can turn to evaluating the climate change response in Part II.

The first step, then, is to define the scope of climate justice. This current chapter argues that both relational and non-relational elements are necessary to fully include the primary victims of climate change in considerations of justice. It will be illustrated that determining what future generations are owed requires non-relational analysis, whereas exploring what is owed to less developed countries requires relational analysis. Furthermore, the chapter seeks to illustrate that relational and non-relational positions are compatible in the case of climate change. In this way, the chapter defends, and begins to develop, a climate justice account that includes both relational and non-relational elements. This account will be further developed in Chapter 2, when defining the non-relational grounds of climate justice, and Chapter 3, when applying the mixed position in order to develop three demands of climate justice. For now, the chapter turns to exploring the apparent divide between relational and non-relational accounts, before setting out reasons for mixing the accounts, and finally defining the mixed position.

COSMOPOLITAN CLIMATE JUSTICE: THE RELATIONAL AND NON-RELATIONAL DIVIDE

Although cosmopolitan climate justice scholars usually agree that climate change presents a case of global injustice, fierce debates exist within the discipline – in particular about who should be at the core of normative concern. These debates stem from, in great part, the distinction between relational and non-relational justice. Climate justice scholars seem to pick one or the other, either implicitly or explicitly.

Relational accounts are grounded in the idea that social relationships and/or political institutions fundamentally alter the relations in which individuals stand, and hence the principles of distributive justice that are appropriate to them (Sangiovanni 2007: 5). Put more simply, what people deserve and why is dependent on the relationships within which they find themselves. Non-relational accounts of cosmopolitan global justice, on the other hand, reject the idea that the scope of justice depends on the relations in which individuals stand (Sangiovanni 2007: 6). Instead, non-relational scholars argue that all humans should be included in the scope of justice because they are human. Put more simply, what people deserve and why does not depend on anything other than their humanity.

Relational and non-relational accounts share the same two basic functions. The first is to provide a scope of justice which determines who will be included in considerations of justice. The second is to explain what is morally required within this scope. Relational accounts set out the scope of justice by defining a relationship, and then explore what is required within the context of this relationship. For example, a relational scholar could determine that there is a relationship between Hugo and Marta because they are married, and could then explore what is morally required of them within the context of this relationship. These requirements depend, in a relational account, on how Hugo and Marta have defined their marriage – the context of their relationship. Non-relational accounts, on the other hand, do not consider relationships when deciding who to include in the scope of justice. In this sense, a non-relational scholar might argue that all married persons have the right to seek a divorce. The scholar does not need to explore the relationship between Marta and Hugo in order to establish this.

Although cosmopolitan non-relational accounts share a wide scope of justice that includes all humans (and sometimes animals), the accounts can differ in terms of what they demand within this scope. Most commonly, non-relational contents are either sufficitarian or egalitarian (Caney 2011: 527). Sufficitarian non-relational theorists defend a clear moral threshold every human is entitled to – for example, basic human rights, or a certain standard of living. Egalitarian non-relational theorists focus on egalitarian distribution of benefits and burdens between the subjects of justice – for example, access to healthcare, which may be dependent on where individual humans (or animals) find themselves on the planet (Caney 2011: 528). In this sense, although all non-relational scholars defend a global scope of justice, they disagree on what is required within this scope of justice. So, a sufficitarian scholar might claim that Marta and Hugo are both entitled to keep 50% of the assets from their marriage after a divorce, while an egalitarian scholar might claim Marta and Hugo should split their assets in the most egalitarian manner possible within their current circumstances.

In order to understand the differences between relational and non-relational accounts, let us consider three examples of existing climate justice accounts. Patrick Hayden, for example, defends a relational climate justice account. He focuses on the relationships created by climate change, and what these imply in terms of duties of justice. Hayden (2010: 358) explains that climate change extends the shared fates and interest of persons beyond political boundaries, creating true global interdependency, and therefore makes the case for a 'globalised' version of John Rawls' conception of justice in order to tackle climate change. Similarly, Steven Vanderheiden (2008: 79) explains that the predicted effects of climate change result in substantial global relationships. He insists that these relationships imply that Rawls' theory of justice can be applied globally in order to reform global institutions so that they can fairly redistribute a basic good, defined by Vanderheiden as the global atmosphere.

By contrast, Simon Caney (2009: 491) defends a non-relational account of climate justice that is based on common humanity. More specifically, Caney grounds his conception of climate justice in three human rights which are predicted to be threatened by climate change: the right to life, the right to food and the right to health (Caney 2010b 166). He concludes that any program of combating climate change should not violate the rights he defines, and therefore sets a minimum moral threshold which cannot be crossed (Caney 2010b: 172). In this sense, Caney presents a sufficitarian non-relational conception of justice which does not find global relationships to be morally relevant for defining the scope and demands of justice. These three examples illustrate the tensions between non-relational and relational accounts of cosmopolitan climate justice, as if a clear either/or choice exists between the positions. The chapter aims to make a case for overcoming this dichotomy and puts forward the idea that a mixed account is preferable in the case of climate change.

THE CASE FOR A MIXED ACCOUNT

The case for a mixed account is primarily motivated by the necessity of including both future generations and those living in less developed countries in considerations of justice. The Introduction of this book explained that these two groups are the primary victims of climate change. The book evaluates the global response to climate change and must therefore be clear about who this response should protect and why. It is only by being clear on this that the book is able to explore whether climate change victims are being adequately protected. Leaving out either future generations or less developed countries from considerations of justice is therefore not possible. The chapter will now explore how relational and non-relational accounts might include the two groups of victims in the scope of justice. This exploration will allow the chapter to illustrate that both relational and non-relational elements are necessary to include those who will be most affected in the scope of climate justice.

Including Future Generations in the Scope of Climate Justice

Although climate change is already occurring and increasingly affecting present generations, a temperature change of 2°C will not occur until 2050–2100 (IPCC 2014a: 14). A rise of 2°C is predicted to bring about 'dangerous' climate change: major ice melting, wildfires, ocean acidification and heatwaves (Hansen et al. 2013: 3). These climatic effects will result in forced migration, place stress on water and food resources and result in the spreading of disease, all of which threaten human life. Because they will not occur until 2050–2100, the victims of these effects will predominantly be those who are not yet alive today. To complicate matters, the main benefit of emissions, namely energy production, is largely consumed by the present generation (Gardiner 2004b: 30). This provides a significant incentive for the current generation to take no action, because members of the current generation may never see the environmental benefits of cutting back on energy use. And yet, deciding what action to take on climate change, a decision that will significantly affect future generations, rests completely with the current generation. In this sense, making a decision on what is owed to future generations is a fundamental aspect of responding to the climate change problem – it allows current generations to make sense of what action is required.

The non-relational account has an advantage over the relational account in including future generations in the scope of justice, and thereby defining what is owed to them. This is because under non-relational accounts, future generations are morally as important as those living today, by virtue of their humanity (Caney 2008: 540). Non-relational accounts, as explained above, do not take global relationships as a starting point, but instead focus on a feature of humanity, such as human rights, which all humans, including future humans, have in common. This implies that a non-relational account must seriously consider what exactly present generations owe to future generations and why. In the case of climate change, a non-relational account might prioritise human rights, which means lowering emissions today in order to ensure that the rights of future generations are protected. In this sense, non-relational accounts are well suited to discussing how to include future generations in the scope of climate justice – because their inclusion is considered fundamentally morally important.

A relational account may be less able to address the question of how much is owed to future generations. Relational scholars rely on existing global relationships in order to set the scope of justice and define what is morally required. It is therefore difficult for relational scholars to claim that present generations have duties of justice towards future generations, since future generations are not part of existing global relationships, as they are not alive and participating in these relationships. This leaves traditional means of establishing a relationship very difficult. For example, there is no chance for a reciprocal relationship between present and future generations because future generations cannot offer present generations any reward for their actions. In addition, determining a causal link between current emissions and future damages is not straightforward, making it arduous to establish a relationship based on harm (Sinnott-Armstrong 2010). Furthermore, future generations may have interests, but cannot express these, as they do not yet exist, and therefore have no bargaining power (Gardiner 2004b: 30). The conditions of a relationship between present and future generations are difficult to pin down. A relational account will therefore have much more difficulty defining what is morally required in the case of future generations, and therefore struggle to include these victims of climate change in the scope of climate justice.

In sum, by restricting the scope of justice to specific global relationships, relational scholars will find it difficult to engage with one of the fundamental distributive problems raised by climate change. This is problematic, because not fully factoring in the importance of future generations would ignore the realities of the climate change problem, since future generations represent the primary victims of climate change. Furthermore, if future generations are not considered morally equal to present generations, then there is little reason to act on climate change until present generations are affected, which could be as long as eighty years from today. Waiting to act until then would intensify and

worsen the effects on future generations as emissions continue to col-
lect in the atmosphere, eventually causing irreversible damages that are
likely to exceed the capacity of natural, managed and human systems
to adapt (IPCC 2014a: 73). Therefore, in the case of future generations,
a non-relational account is superior for including one of the primary
groups of climate change within the scope of climate justice.

Including Less Developed Countries in the Scope of Climate Justice

Although the non-relational account has an advantage in the case of
future generations, it is at a disadvantage when it comes to including
the second group of primary victims of climate change, less developed
countries, in the scope of justice. This is especially true in the case of
sufficitarian accounts, which set a global scope of justice, but offer limited guidance on the demands of justice. Often, these scholars simply define a moral threshold that cannot be crossed and do not specify
demands of justice beyond this. An example of this can be found in
Caney's account. He sets a non-relational scope of climate justice, the
demands of which are to not violate the human right to health, life and
sustenance. It is unclear what justice might demand beyond not crossing this moral threshold in the case of climate change. Mathias Risse
has alluded to this problem facing sufficitarian non-relational accounts.
According to Risse (2012:30), simply listing rights makes it difficult to
assess not only how imposing duties are, but also precisely what the
content of duties is. Risse (2012: 8) claims that protecting the rights
established by sufficitarian non-relational scholars requires references
to associations (or relationships) for specific assignments of duties. In
other words, Risse claims that a sufficitarian non-relational account, by
identifying basic rights (or another moral threshold), only offers a very
limited insight into what justice might demand.

This problem with non-relational accounts is apparent when considering how to include those living in less developed countries in the
scope of justice. As was explained in the Introduction of this book, the
impacts of climate change will fall disproportionately upon developing
countries. In fact, the World Bank (2010: xx) estimates that less developed countries will bear 75–80% of the burdens of climate change. This
is for two broad reasons. First, less developed countries are located in
areas that will be hardest hit by climate change effects (IPCC 2007: 12).
Second, less developed countries often have other problems, such as
poverty or weak infrastructure, which create conditions of low adaptive

capacity to climate change (IPCC 2007: 12). These countries may therefore not have the financial capacity or infrastructure necessary to combat the ill effects of climate change, and will as a consequence not be able to prepare or defend themselves against effects like flooding, droughts or rising sea levels as effectively as developed countries.

These realities of the climate change problem create a relationship between developed and less developed countries, because developed countries are engaging in behaviour which endangers the people living in less developed countries. To complicate matters, some less developed countries are now developing to the point where their emissions are as high as those of developed countries. China, for example, is now the world's largest emitter in absolute terms and also ranks on a par with the European Union in per capita terms. This raises profound questions of fairness, for example on how much developed countries owe to less developed countries, and whether some less developed countries may need to contribute to the global climate change reduction efforts due to their level of emissions and/or wealth. These questions are by their nature questions of distributive justice, because they concern what is owed and what is deserved, or more specifically how benefits and burdens should be distributed. A climate justice account must be able to answer these distributive questions in order to determine what is owed to whom, and who should be responsible for what climate change action. These are fundamental normative questions raised by the climate change problem. It is especially important to answer these questions when assessing the global response to climate change, because assessing the response requires clarity of what is expected of policymakers and why.

The distributive questions outlined above are not merely theoretical. Global negotiations under the United Nations Framework Convention on Climate Change (UNFCCC) have revealed that less developed countries are very concerned with the idea that developed countries are historically at fault. These concerns are reflected in the Convention on Climate Change, which states that 'the largest share of historical and current global emissions of greenhouse gases has originated in developed countries' (UNFCCC 1992: 2). As a result of this historical responsibility, the UNFCCC (1992: 4) calls exclusively on developed countries to 'take immediate action'. This prioritisation of developed countries has not been uncontroversial, and yet it to this day presents a cornerstone of multilateral climate change action. The Paris Agreement states in Article 4.4 that 'developed country Parties should continue taking the lead by undertaking economy wide absolute emission reduction targets' (UNFCCC 2015: 22). However, the Paris Agreement also notes, in the same Article, that 'developing country Parties should continue enhancing their mitigation efforts and are encouraged to move over time towards economy wide emission reduction or limitation targets in the light of different national circumstances' (UNFCCC 2015: 22). It is clear to see from this somewhat vague wording that who is responsible for how much and when is still up for debate, implying that there remain distributive justice questions that must be answered. A relational account has an advantage in answering these questions because it is better suited to understanding and assessing relationships.

To illustrate the advantages of the relational account, consider what non-relational accounts might have to say about the relationship between developed and less developed countries. A sufficitarian non-relational account could assert that there is a duty to stop violating certain human rights, which could be used to make the case for lowered emissions by developed countries in order to protect the rights of people living in less developed countries. However, such an account is not able incorporate the nuances of the relationship between less developed and developed countries beyond these minimal demands. As David Miller (2013: 5) explains, the relationship in which parties stand to one another must be properly understood before one can say what justice requires these parties to do. The relationship between less developed and developed countries involves more than the problem of human rights violations, because it raises questions about historical responsibility, what is owed to less developed countries, and how much high-emitting but less developed countries must engage in mitigating climate change. A sufficitarian account would find it difficult to engage with these nuances because it does not focus on relationships in order to define demands of justice.

To simplify this idea further, consider the following non-climate-based scenario. Imagine that five people want to share a cake. A non-relational sufficitarian account would require that all five individuals get a certain amount of cake which is in line with the sufficitarian moral minimum defined. For example, sufficitarians who defend a right to life may say that each person is entitled to a number of calories which will keep them alive. However, if the cake has more calories than are required to keep five individuals alive, then it is unclear what each person is entitled to. Under a sufficitarian account, individual A, who is greedy and arrogant, and believes that she deserves more, could reasonably take the biggest piece by force, leaving four pieces which each have enough calories for the remaining individuals. This would be just under a non-relational sufficitarian account, because the basic rights of the other four individuals would not be violated. Applied to the climate change case, this implies developed countries could take more of the emissions share than less developed countries, as long as the right to life of individuals in less developed countries is protected. However, this ignores complexities like historical fault, level of development or level of wealth, all of which raise questions about distribution. In sum, a sufficitarian non-relational account seems unable to capture the complexities of the relationship between developed and less developed countries.

Caney offers a response to this type of criticism. He claims that egalitarian non-relational accounts can accommodate the thought that increased ties (relationships) have normative implications to people's entitlements of distributive justice. Caney (2011: 526) explains this in the following way: 'although humanity-centred cosmopolitan egalitarians hold that some distributive principles apply independently of persons' membership of a common association, the substantive implications of those principles will be affected by the extent to which persons belong to a common association and the extent to which that association is coercive and characterised by high levels of interdependence'. What Caney is implying here is that the distribution of benefits and burdens will depend on the context of existing relationships. Caney uses the example of the right to health to illustrate his point. He explains that although every human is entitled to the highest attainable standard of healthcare, the standard which persons receive will depend on where they live (Caney 2011: 526). If a community is, for example, isolated and has no roads by which medicines can be delivered, then the highest standard of healthcare this community can receive is lower than the standard which can be received in other, more connected, communities (Caney 2011: 527). As the isolated community becomes more connected, the standard of healthcare they receive will improve. The point Caney (2011: 527) is making is that 'what justice means in practice will vary as interdependence increases'.

According to Caney (2011: 529), accepting this implies that egalitarian accounts of non-relational justice are able to capture the ways in which existing relationships have moral relevance for distributive justice. However, it is questionable whether an egalitarian account could truly capture the complexities of the relationship between developed and less developed countries, because this relationship raises distributive questions that cannot be fully addressed by demanding that there should be an egalitarian distribution. Answering these distributive questions will require a thorough exploration of the complexities of the relationship between developed and less developed countries and outlining demands of justice that are specific to this relationship, which requires relational analysis. Taking relationships as a starting point for grounding demands of justice means that the complexities of those relationships can be addressed, and answers to the distributive questions raised by a relationship can form a constitutive part of the climate justice account defended here, rather than something which is considered after the demands have already been defined.

To illustrate this more simply, consider the cake scenario again. Under an egalitarian account, the five individuals would not merely be entitled to enough calories to stay alive, but to an egalitarian distribution of the pieces of cake. However, without discussing the relationships between the five individuals eating cake, it would be difficult to define what is required for egalitarian distribution. Caney claims that the moral relevance of relationships is taken into account in the egalitarian position, in the sense that the account can acknowledge that relationships will affect distribution. An egalitarian account could on this logic acknowledge that the greedy and arrogant individual A will take more cake, and that this will affect distribution, just as the access to health care affects the distribution of the right to health. However, without considering the relationship between individual A and the other four individuals, this simple acknowledgement can potentially ignore the realities of the relationship that raise important distributive questions. For example, individual A may be a serial cake-stealer, and never allow the other four individuals a bigger piece, even though these four individuals bought the ingredients for the cake and baked it. Surely this should be factored in when deciding on the demands of justice, rather than acknowledged as affecting distribution.

This may seem slightly spurious, but in the case of climate change, developed countries have historically taken the biggest share of emissions, which is now causing climate change, and which will harm persons in less developed countries who have not contributed to global emissions. In addition, some less developed countries are now developing at a rate which has placed them into the top ten of highest-emitting countries. This raises distributive questions about how much developed countries owe less developed countries, and whether less developed countries have any moral responsibility to address climate change when they have so far not contributed to the problem on the same scale as developed countries. An egalitarian non-relational model that simply calls for an egalitarian distribution evades these questions by claiming that the relationship may affect distribution and not discussing these effects when defining the demands of climate justice.

A relational account, by contrast, would take the distributive questions raised by the relationship between developed and less developed countries as a starting point for formulating what climate justice demands. This is because relational accounts are based on the idea that principles of distributive justice cannot be formulated or justified independently of the practices they are intended to regulate (Sangiovanni 2007: 5). In the cake scenario, a relational account would take the relationship between the five individuals as a starting point and explore what is owed to whom on the basis of this relationship. The relational account could, for example, discuss whether the four individuals who bought the ingredients for the cake may deserve more than individual A, or whether Individual A, due to her past behaviour, may owe something to the others. In the case of climate change, a relational account could take the relationship between less developed and developed countries as a starting point and address whether historical fault has implications on responsibility, or whether level of emissions, rather than level of development, has a bearing on how much countries must contribute to global mitigation efforts. The relational account could then formulate demands of justice which serve to answer these questions directly, instead of merely acknowledging their moral significance, as in Caney's egalitarian account.

In sum, what is being put forward here is that a normative understanding and assessment of the distributive questions raised by the relationship between developed and less developed countries requires a relational account, because this account formulates demands of justice which are based on the specifics of relationships. On this logic, a relational account must form part of the climate justice account defended here, because the book aims to include less developed countries in the scope of justice. Furthermore, considering the fact that every person on the planet has the potential to contribute to the climate change problem and will potentially be affected by its consequences, climate change will no doubt create other relationships which raise new ethical or distributive questions that may not have previously been addressed. For example, to what extent should individuals be held to account for their emissions? Should corporations pay for adaptation costs? A relational account is especially useful for such questions. Indeed, Chapter 3 will use relational analysis in order to define the duties of non-state actors. But first, the mixed position must be defended below.

DEFENDING THE MIXED ACCOUNT

In order to defend a climate justice position that includes both relational and non-relational elements, one must consider whether either account should be prioritised over the other. To make this decision, the chapter will consider which side of the account would form a stronger normative foundation for a climate justice account. The Introduction of the book explained that cosmopolitans take the equal moral worth of humans as their starting point. Some cosmopolitans argue that defining justice as a condition of a special relationship, as relational justice scholars do, seems to contradict this cosmopolitan premise. Tan Kok-Chor (2004: 59), for example, argues that 'constraining the applicability of justice to whatever social arrangements we currently happen to have would arbitrarily favour the status quo, which is plainly contrary to the aims of justice'. Furthermore, relational justice implies that it is only when humans enter into relationships that distributive justice is required. This appears incompatible with the cosmopolitan assumption that all humans are morally equal. The moral worth of humans is not defined as conditional by cosmopolitans. It therefore seems that the non-relational account supports cosmopolitan aims more fully than the relational account.

This is an important advantage of non-relational accounts in terms of cosmopolitan consistency. The book therefore puts forward that the non-relational account is most suited to forming a foundation for the mixed position. If the mixed position is based on a scope of justice which is in line with the fundamentals of cosmopolitanism, this will make for strong normative underpinning. For this reason, the mixed account will use the non-relational side of the account as a foundation that defines a global scope and establishes a minimal moral threshold, and then use the relational side of the account to define demands of justice based on the relationships created by climate change. In this sense, climate relationships will form a constitutive part of the mixed climate change justice position, because they will be used to define relationship-specific demands of justice in Chapter 3. These demands will be underwritten by a non-relational baseline, defined in Chapter 2, which forms the immutable foundation of the mixed account.

As this is the first attempt to develop a mixed climate justice account, it may be helpful to explain how the proposed account differs from David Miller's split-level account, which is an existing (but not climate-change-based) mixed account. Explaining how the cosmopolitan account defended here differs from Miller's will help to further clarify the proposed mixed position. At first glance, this account seems similar to the mixed position defended here. Under Miller's model, basic rights must be respected at the global level, which is a non-relational demand; but more complex demands of justice are defined within states, which suggests a relational account, because states represent a specific relationship within which demands of justice are defined. Miller's account is therefore a mixed account that combines non-relational and relational elements. However, the mixed account presented here very much differs from Miller's account because Miller is not consistent in his non-relational defence of basic rights at the global level. For example, Miller (2005: 74) argues that according to his split-level account, a nation could offer international aid if it meant minimally reducing national education funding, but not if it meant that fellow nationals would starve. Here Miller is attempting to explain that relationships to fellow nationals can override non-relational human rights of those outside national borders. In his example, starving nationals would be prioritised over international aid that would help starving strangers abroad. In other words, the human rights of non-nationals can be overridden by the human rights of fellow nationals. Or, to put it another way, the non-relational demands Miller sets out can be overridden by relational duties.

This is in stark contrast to what is being posited here. The non-relational element of the mixed account sketched above ensures that non-relational demands of climate justice can *under no circumstances* be disregarded. The scope of justice is immutably non-relational, and the relational element merely helps to define more exact demands of justice, rather than overriding the non-relational scope of justice. The rationale behind this can be found in the basis of cosmopolitan theory: the equal moral status of humans. In the mixed account defended here, all humans, by virtue of their humanity, are included in the scope of justice, and cannot be removed from the scope of justice because of a special relationship. The non-relational scope sets a moral minimum that must never be crossed. This ensures that the mixed account is consistent as a cosmopolitan position.

Prioritisation of the non relational account is important not only for cosmopolitan consistency, but also in order to answer some common concerns associated with relational cosmopolitan accounts. The first concern is the idea that morally arbitrary factors should not affect what people deserve under a cosmopolitan account. Caney (2005: 111), for example, claims that 'it is difficult to see how and why the fact that one group of people is linked by interaction should impact on their entitlements'. This is a valid criticism of relational accounts, because if the scope of distributive justice is limited to certain social schemes, then this will affect what people are entitled to. However, in the mixed account outlined here, every human being, including future generations, is included in the scope of justice. In addition, the relational side of the account is not used in order to explore morally arbitrary circumstances, but rather to discuss specific distributive justice questions that arise, for example, from the relationship between developed and less developed countries. Wealth and emissions levels, both past and present, are not morally arbitrary, because these factors impact on what is considered fair and just to the parties within an existing relationship and are therefore important moral considerations.

A second concern is that individuals could fall outside the scope of justice if they are not part of the relationships chosen by relational scholars, or if they happen to leave these relationships at a later time. This is problematic, because individuals who fall outside of the scope of justice would not be considered full moral equals of those within the relationship. In this way, these individuals would not share the status of being a primary unit of moral concern and would not be considered to possess equal moral worth to those who remain in the relationship. This is not acceptable from a cosmopolitan perspective, which emphasises the equal moral status of all individuals. Fortunately, the mixed account defended here can overcome this criticism. In the mixed account, there is a non-relational minimum that can never be crossed, and which ensures that no individual is excluded from the scope of distributive justice. Moreover, this non-relational minimum implies that individuals are the primary unit of moral concern and possess equal moral worth. Those individuals who participate in the relationship of developed and less developed countries are entitled to the same moral concern as every other individual, and are not entitled to additional privileges. Instead, those participating in relationships may be responsible for lowering emissions or they may be owed assistance in adapting to climate change, depending on the circumstance. This does not afford these individuals any additional privileges, but rather responds to the questions of distributive justice that exist in the case of climate change, and which must be answered.

Now that the mixed account has been set out and defended, the book turns to applying the account in Chapters 2 and 3. It is worth stressing that there has so far been no explicit attempt to mix non-relational and relational justice in the climate change justice literature. As a result, there may be some scepticism as to whether this mixed account is defensible. Chapters 2 and 3 will continue the discussion on how to apply a mixed account, hopefully convincing sceptics that it is indeed plausible and useful to mix these approaches in the case of climate change.

CONCLUSION

This chapter put forward that relational and non-relational elements are necessary to fully include the primary victims of climate change in the scope of climate justice. It was argued that determining what future generations are owed requires non-relational analysis, whereas exploring what is owed to less developed countries requires relational analysis. The chapter therefore made the case for a position that is both relational and non-relational. This mixed account is based, in the first instance, on a non-relational scope that encompasses every individual, including future individuals. The relational element of the account is then applied in order to explore existing climate change relationships – for example, the relationship between developed and less developed countries. This allows the mixed account to put forward demands of justice that stem from relational analysis. In this sense, the mixed account uses the non-relational side of the account as a foundation that defines a global scope and establishes a minimal moral threshold, and the relational side of the account to define demands of justice based on the relationships created by climate change. This mixed account will now be further developed in Chapter 2, which concerns defining the grounds of the non-relational side of the account.

Chapter 2

THE GROUNDS OF CLIMATE JUSTICE

INTRODUCTION

This book aims to evaluate the global response to climate change. An important part of this evaluation is the development of a climate justice account. One cannot assess what is going right and wrong if one does not define what the right, or rather just, response to climate change would look like in the first place. Who will be most affected, what exactly is at stake, what action must be taken in the face of climate change, and who should be responsible for this action are all important questions that must be answered. The previous chapter focused on the first question – who will be most affected – and argued that the scope of justice should be relational and non-relational in order to include the primary victims of climate change in the realm of moral concern. This current chapter now turns to the second question – what exactly is at stake – by setting out the grounds of climate justice. Defining the grounds of justice is a key task for any climate justice account because it allows readers to understand what must be normatively prioritised. The grounds of justice in this sense represent the moral underpinnings of the climate justice account, a normative subfloor that must not be crossed. The final two questions – what action must be taken in the face of climate change, and who should be responsible for this action – will be the subject of Chapter 3, which sets out three demands of climate justice in order to clarify what exactly is expected from the global response to climate change.

The previous chapter defended a mixed scope of justice that has both relational and non-relational elements. It was explained that the non-relational side of the account serves to set a minimum moral threshold that applies to all individuals, no matter the time or place of birth. The book must therefore now define this minimal moral threshold, which will constitute the grounds of climate justice: the human right to health. Health experts are confident that climate change will contribute to the global burden of disease and premature death, and have deleterious consequences for human health (Hansen et al. 2013: 8). And yet, despite the threats climate change poses to human health, health has been somewhat neglected among climate justice scholars and policymakers alike. This chapter aims to illustrate the importance of human health in the case of climate change. Furthermore, the book has the wider aim of demonstrating that a focus on human health can reveal much about the inadequacies of the global response to climate change.

For now, the book will focus on defending the use of the human right to health as a minimal moral threshold, or grounds of climate justice. This defence will unfold as follows. First, this chapter will out-line the reasons behind using a human-rights-based approach. Next, it will turn to discussing why the human right to health, in particular, should be prioritised over other human rights that are threatened by climate change, including the rights to water, food and free movement. Finally, it will provide a definition of the right to health and begin to explain what action is necessary in order to protect this right. This will be expanded on in Chapter 3.

TAKING A HUMAN RIGHTS APPROACH

This book takes a human rights approach to defining a minimum moral threshold that cannot be crossed, referred to here as the grounds of justice, for the following reasons. First, human rights, in political phi-losophy, commonly represent moral thresholds. They designate the most basic moral standards to which persons are entitled and specify the line beneath which no one is allowed to sink (Caney 2010b: 164). If a human rights approach is commonly recognised as providing a moral threshold by other cosmopolitan scholars and political philoso-phers more broadly, then using such an approach will be recognised as valid and familiar by fellow scholars. This makes the task of defining a moral threshold slightly more straightforward, because a human rights approach is not seen as controversial or previously untested. Taking a human rights approach therefore comes with the advantage of having been previously defended and established by global justice scholars.

Second, a human rights approach can 'humanise' climate change. Climate change is often seen as a problem for nature, or animals – the overused image of a polar bear floating on ice to represent the effects of climate change comes to mind. The human aspect of climate change is all too often forgotten both by policymakers and individuals around the world more widely. A human rights lens can focus international climate debates on the effects of climate change on human lives, which may add a sense of urgency and help policymakers take existing failures more seriously (Limon 2009: 451). If the object of climate policy is to prevent severe human suffering rather than environmental phenomena, it may become more difficult for governments to avoid action (Harris 2013: 135). In this sense, individuals, scholars and policymakers alike may feel a greater sense of urgency to protect the human right to health than to lower emissions for more abstract reasons, such as methane levels in the atmosphere causing environmental damage.

Third and relatedly, human rights are recognisable as normatively important to policymakers. Designating a human right (the right to health) to protect is therefore a worthwhile task. Using a human rights approach could refine the assessment of the human costs of climate change, mobilising substantial adaptation funding and increasing the likelihood of setting sound mitigation targets (ICHR 2008: vii). Furthermore, focusing on human rights could enable policymakers to understand and take account of the needs of the most vulnerable, potentially increasing urgency to act on climate change (ICHR 2008: 7). As the International Council on Human Rights, or ICHR (2008: 7), explains, understanding human rights implications could therefore play a valuable role during climate change negotiations. The discussion of the human right to health in this chapter, and the assessment of how well this is protected by policymakers in Part II of the book, will therefore no doubt be of value within and outside of the academic community.

Fourth, climate change is increasingly recognised as a human rights concern by global governance actors, which this book concerns itself with assessing. The UN highlighted the impact of climate on human rights in its 2008 Human Development Report (2007: 4), noting that 'climate change represents a systematic violation of the human rights of the world's poor and future generations, and a step back from universal values'. Furthermore, in 2007, the deputy UN Human Rights Commissioner Kyung-wha Kang explained that 'any strategy to deal with climate change, whether in terms of adaptation or mitigation, must incorporate the consequences for humans, as individuals and communities, and the human rights framework is the most effective way to do so' (Kang 2007: 2). In this sense, the connection between human rights and climate change is becoming increasingly established on the global stage. Assessing to what extent the climate response protects human rights will therefore no doubt be of interest to global governance actors.

Fifth and finally, human-rights-based approaches are not uncommon within climate justice literature. In this sense, the book's approach is part of an emerging trend, and speaks to existing accounts of climate justice. Engaging with existing literature, and building on this, is an important task for this book, since it aims to illustrate the value of assessing practice to those in the field of climate justice. Interestingly, some climate justice scholars define environmental human rights while others work with existing human rights, as this book does. Patrick Hayden (2010: 361), for example, defends the right to be protected from environmental harm, which encompasses the right to clean air and water. Similarly, Tim Hayward (2007: 432) defends the human right to ecological space, defined as the human right to live in an environment free of harmful pollution. In contrast, Simon Caney (2010b: 168) bases his conception of justice on human rights previously defined in the Universal Declaration of Human Rights: the right to life, health and sustenance.

Like Caney, this book will defend an existing human right – the human right to health. The right to health can be found in Article Twenty-Five of the Universal Declaration of Human Rights (UDHR). Although the UDHR is not a legal document, it is a well-known and well-referenced list of human rights. There is no question that human rights enshrined here have attained the status of a lingua franca of global moral discourse (Beitz and Goodin 2009: 1). If the climate justice account defended in this book is to impact climate change policy, it may be best to work with rights that have previously been established as morally important and that are recognisable as such by those working within global climate change governance.

While there have been attempts to define a comprehensive list of climate-change-related rights (for example, in the *Draft Declaration of Principles on Human Rights and Environment*), these rights have not become part of the normative discourse in the same manner as existing human rights. In addition, the Draft Declaration focuses on 'the linkage between human rights and the environment' rather than establishing new environmental rights (Hayden 2001: 670). The same can be said of those who defend environmental rights within cosmopolitan literature. Hayward's conception of the human right to ecological space, for example, seeks to ensure the right of each individual to an environment adequate for their health and well-being (Hayward 2007: 440).

This definition directly incorporates the definition of the right to health found in the Universal Declaration of Human Rights. Hayward is therefore defending an existing human right under new environmental language. This indicates that environmental rights often encompass or rename existing rights, both among policymakers and scholars. It is therefore questionable whether separate environmental rights present a meaningful addition. Defining a right specific to the climate change problem seems to be a time-consuming and potentially unnecessary in-between step, when it may be more expedient and impactful to discuss well-established human rights threatened by climate change. This is particularly true when considering that human rights may spur on policymakers, as discussed above. For all of these reasons, the book prioritises a previously recognised human right, namely the human right to health, rather than defining new climate change rights.[1]

PRIORITISING THE HUMAN RIGHT TO HEALTH

The main purpose of the book is to conduct a cosmopolitan assessment of the global response to climate change. It would be impossible to conduct such an assessment without a clearly defined climate justice account. What is expected of policymakers in multilateral and transnational climate change governance must be set out in a comprehensive manner. Otherwise, it will be very challenging to analyse to what extent policymakers have met these expectations with any great precision. An important part of setting out what is normatively expected of policymakers is to explain what must be prioritised and protected, or in other words, defining what is at stake. This will determine what action is required and who might be responsible for taking this action. By prioritising one human right, namely the human right to health, the chapter is able to set out a very precise moral threshold which in turn can be used to define exact demands of justice. These demands can then be used in the assessment of the climate change response in Part II of the book.

It is important to acknowledge that climate change does not exclusively threaten the human right to health. There are other rights at stake, including the right to food and water, the right to life and the right to free movement, all of which will be discussed below. In this sense, prioritising the human right to health does not mean denying the importance of other human rights. Instead, the book aims to draw attention to a right which has been neglected by scholars and policymakers alike, despite its importance to the climate change problem. In addition, the book aims to illustrate why the human right to health, in particular, can provide a very strong foundation for a climate justice account. As will be explained below, the human right to health can capture the threats climate change poses to humans more comprehensively than other key human rights.

To illustrate why this is the case, the chapter will now outline the threats climate change poses to humans: food security, water shortages, displacement and health (IPCC 2014a: 13–16). All of these risks directly threaten humans and their human rights. Let us begin by considering food security. With global population expected to rise, and food demand predicted to increase exponentially, climate change is expected to pose large risks to food security globally for a number of reasons (IPCC 2014a: 13). For one, climate change is expected to affect global marine species, undermining fishery productivity (IPCC 2014a: 13). In addition, climate change will affect rice and maize production, particularly in tropical and temperate regions (IPCC 2014a: 13). Furthermore, climate change is projected to reduce renewable surface water and groundwater resources in most dry subtropical regions (IPCC 2014a: 13). This will threaten crop growth and agricultural outputs, resulting in food shortages. Moreover, the reduction of surface and groundwater threatens clean water supplies. This results in a second risk, water shortages. These shortages may lead to increased thirst and dehydration-related deaths, and furthermore potentially increase competition for water, causing conflict (IPCC 2014a: 13).

Although the Universal Deceleration of Human Rights does not define a right to food and water, threats to these basic human needs for sustenance can be linked to other human rights, including the right to 'life, liberty and security of persons' defined in Article One. Lack of food and water are no doubt life-threatening, putting the right to life at risk. The World Health Organization predicts that climate change is expected to cause approximately 250,000 additional deaths per year between 2030 and 2050 (WHO 2017). In this sense, it is obvious that climate change threatens the human right to life. In addition, security of persons will be severely affected should there be conflict over resources, as is currently predicted (IPCC 2014a: 16). Climate change is the ultimate 'threat multiplier' – aggravating already fragile situations and potentially contributing to further social tensions and upheaval (UNEP 2017). *A New Climate for Peace: Taking Action on Climate and Fragility Risks*, an independent report commissioned by the G7, identifies several climate-fragility risks that pose serious threats to the stability of states and societies in the decades ahead (G7 2015). These include resource competition, volatile food prices and provision, and transboundary water management. The threat to security of persons is already well on the way to being established by policy experts.

The third threat, displacement, is projected to increase as a result of climate change (IPCC 2014a: 16). If crops fail, water supplies run short and land becomes uninhabitable, it will be impossible for populations to remain. In addition, some territories will simply disappear due to sea-level rise, most notably small island states. This will no doubt result in displacement of entire populations. Furthermore, extreme weather events such as storm surges and tropical cyclones are predicted to lead to mass displacements (Biermann and Boas 2010: 68). It is predicted that most refugees will come from the continents of Africa and Asia. Both regions are at high risk of extreme weather events and sea-level rise, and will also be severely affected by drought and water scarcity (Biermann and Boas 2010: 69). Although the exact numbers of climate refugees vary from assessment to assessment depending on methods, scenarios, timeframes and assumptions, the available evidence indicates that the climate refugee crisis will surpass all known refugee crises in terms of the number of people affected (Biermann and Boas 2010: 61). Myers (2002: 609), for example, estimates that the total number of people at risk from sea-level rise by 2050 is likely to be 162 million. In addition, 50 million people could become refugees due to droughts, floods and extreme weather events (Meyers 2002: 609). In total, then, Myers expects 212 million climate refugees by 2050. Compare this to the Syrian refugee crisis, where 13.5 million are currently in need of humanitarian assistance (UNHCR 2017). Meyer's predictions are in line with most estimates: 200–250 million climate refugees by 2050 are expected according to a number of studies (Biermann and Boas 2010: 68).

There is no human right to not be displaced, but Article Thirteen of the Universal Declaration of Human Rights specifies that 'everyone has the right to freedom of movement and residence within the borders of each state' and 'everyone has the right to return to his country'. Climate change threatens both of these human rights, because there may not be a choice to move freely, or indeed to ever return home, especially in the case of small island states. The human rights of refugees are also protected by the United Nations High Commissioner for Refugees (UNHCR), which is already working on strengthening the resilience of people who are particularly vulnerable to the effects of climate change

(UNHCR 2017). The UNHCR is the primary global governance actor responsible for refugees and has explained that 'the consequences of climate change are enormous' for their agency (UNHCR 2017). All of this strongly indicates that the human rights to freedom of movement, residency and right to return home are very much predicted to be under threat by climate change.

Finally, there is a substantial risk to health. The risks to health are multifaceted, encompassing both direct and indirect threats. Heat stress and air pollution, for example, both pose a direct threat to human health because they can cause or aggravate cardiovascular and respiratory disease, particularly among elderly people (WHO 2017). Flooding, landslides and extreme precipitation pose another direct threat because they can affect clean water supplies and thereby spread diseases, including diarrheal diseases (IPCC 2014a: 15). Floods can also increase diseases transmitted through insects, snails or other cold-blooded animals, as well as mosquitos, because they create breeding grounds for these disease-carrying species. Floods also cause drowning and physical injury, damage homes and disrupt the supply of medical and health services, all of which threatens human health (WHO 2017).

Threats to health are also indirect. For example, droughts, inland and coastal flooding, extreme precipitation and sea-level rises all threaten crop growth and water supply (Hansen et al. 2013: 8). Stress on these resources affects health because it can lead to malnutrition, resulting in various mineral and nutrient deficiencies. Nutrient deficiencies can lead to scurvy, anaemia, bone loss and high blood pressure, among other health problems. Malnutrition can also increase susceptibility to and severity of infections and is thus a major component of illness and death from disease (Müller and Krawinkel 2002). In addition, lack of food and water inhibits the ability to provide nutritionally balanced diets to mitigate and repair liquid and nutrient losses when diseases develop, potentially increasing their morbidity rates (Jamison et al. 2006: 373). Other indirect threats to health include increased instances of existing diseases such as malaria, because warmer weather spawns more mosquitos, as well as a rise in infectious diseases including HIV and AIDS, as more and more people become displaced (Caney 2010b: 167). Forced migration may also lead to people living in crowded and unsanitary conditions. This threatens health because communicable diseases are exacerbated by poor housing, crowding, dirt floors, lack of access to sufficient clean water or to sanitary disposal of faecal waste, and a lack of refrigerated storage for food (Jamison et al. 2006: 373).

As the WHO (2017) explains, a lack of safe water can compromise hygiene and increase the risk of diarrhoeal disease. In addition, rising sea levels and increasingly extreme weather events will destroy homes, medical facilities and other essential services (WHO 2017).

From the above, it is clear that health is not only threatened by disease or changes in temperature, but also by food and water shortages as well as displacement. The threat to health is seemingly related to and intertwined with the three threats previously discussed: food, water and displacement. In fact, the WHO (2017) describes the threat from climate change as a threat to the social and environmental determinants of health, including clean air, safe drinking water, sufficient food and secure shelter. In this sense, the WHO is very much aware that health is interlinked very strongly with other climate change threats. This is why the right to health is so important in the case of climate change. In a sense, health helps to capture the complexity of the way in which climate change threatens humans. Consider, as an illustration, the definition of the right to health provided by the UDHR in Article Twenty-Five, which includes health, food and shelter:

> Everyone has the right to a standard of living adequate for the health and well-being of himself and of his family, including food, clothing, housing and medical care and necessary social services, and the right to security in the event of unemployment, sickness, disability, widowhood, old age or other lack of livelihood in circumstances beyond his control.

Food and shelter both present a constitutive condition of human health according to this definition. In this sense, the right to health captures much more than may initially be assumed. If the right to health is compared to the definition of the right to life, or the right to free movement, it is much more far-reaching and encompassing than these rights. Grounding the climate justice account in the right to life, for example, would ignore the multifaceted nature of climate change threats. Climate change will not merely kill humans or put their security at risk, it will also make them ill, force them to flee and create water and food shortages that would further exacerbate illnesses. The right to 'life, liberty and security of persons' does not capture these complexities. Similarly, grounding the climate justice account in the right to free movement and to return home would also be too narrow. The right to free movement is no doubt multifaceted because displacement is caused by a wide variety of climate change effects, including sea-level rise, drought, floods and extreme weather events. However, grounding the climate justice position in the right to free movement would fail to capture the threats to those who are not displaced, including malnutrition, conflict and the spread of disease, among others. The right to health, by comparison, is able to capture what happens to those who stay where they are and to those who must flee, as well as capturing the effects of food and water shortages to a greater extent that the right to 'life, liberty and security of persons'. Staying alive and secure are of course normatively important but present a very low threshold that fails to capture the dire effects climate change will have on human health for those who survive.

So, although there are no doubt other human rights that are threatened by climate change, the right to health is the most encompassing of the multifaceted threats to humans that climate change presents. This perception of the right to health is reflected in recent scientific research. James Hansen et al. (2013: 8), for example, explain that food shortage, polluted air, contaminated or scarce supplies of water and an expanding area of vectors causing infectious diseases are all threats to human health. For this reason, Hansen et al. prioritise the risk to health in their discussion of what should be done to protect young people from climate change. The book is therefore not alone in its prioritisation of human health, although other cosmopolitans have not yet done so. In sum, what has been argued above is that the right to health provides a very strong and multifaceted normative foundation to the climate justice account, because of its far reach and ability to capture the complexities of the threat climate change poses to humans. It is for this reason that the book prioritises the right to health as the grounds of the climate justice position.

DEFINING THE HUMAN RIGHT TO HEALTH

Now that the reasons behind prioritising the right to health have been outlined, this chapter turns to developing a definition of the right to health that can be used to ground the climate justice account. As was argued above, rights that are part of existing global normative discourse are particularly useful for a conception of climate justice. It is therefore important to point out that the right to health has been defined in various internationally recognised documents. Aside from being enshrined in the Universal Declaration of Human Rights, the WHO (1946: 1) has defended the 'enjoyment of the highest attainable standard of health' as a fundamental human right. Furthermore, the right to health is affirmed in several regional conventions, including the 1948 American Declaration of the Rights and Duties of Man, the 1981 African Charter on Human and Peoples' Rights and the 2000 Charter of Fundamental Rights of the European Union. Recently, scholars including Patrick Hayden (2012: 571) have gone so far as to claim that the human right to health has surged onto the international stage as one of the most pressing human rights of the twenty-first century. Therefore, it can be said that a right to health is well established in global normative discourse.

The right to health is often defined in a broad and far-reaching manner. For example, the International Covenant on Economic, Social and Cultural Rights defines the right to health as the 'right of everyone to the enjoyment of the highest attainable standard of physical and mental health'. In this way, the right to health is not understood as a right to be 'healthy'. Rather, it is defined as a right to the enjoyment of a variety of diagnostic, curative and preventive facilities, goods, services and conditions necessary for the realisation of the highest attainable standard of health (Hayden 2012: 572). Similarly, the definition provided by Article Twenty-Five of the UDHR is very broad:

> Everyone has the right to a standard of living adequate for the health and well-being of himself and of his family, including food, clothing, housing and medical care and necessary social services, and the right to security in the event of unemployment, sickness, disability, widowhood, old age or other lack of livelihood in circumstances beyond his control.

Although these broad definitions are no doubt valuable because health is a complex matter, such broad conceptions make it nearly impossible to define a minimal moral threshold. The 'highest attainable standard of health', for example, is necessarily context-specific and includes many possible elements such as mental health, family planning, vaccinations, nutrition and other factors, which means a definition is difficult to pin down. Furthermore, including aspects such as 'social services and the right to security in case of unemployment', as in Article Twenty-Five above, would make the expectations of policymakers difficult to pin down. Agreeing on what 'well-being' means to the primary victims of climate change would also be nearly impossible. Less developed countries encompass billions of people across several continents, and their ideas on what 'well-being' might mean will surely be starkly contrasting and differentiated. Determining what well-being means to future generations is equally difficult, since we cannot ask them what makes them feel 'well'. In short, broad approaches such as the ones above would make it very difficult to set out requirements for policymakers, hampering the possibility of assessing current practice with any great precision. Using a far more minimal definition of the right to health would allow for a modest yet clear moral minimum in the case of climate change. In addition, it is valuable to examine to what extent a very basic, or minimal, conception of a right to health has been protected by multilateral and transnational actors, because this will be more revealing about the inadequacies of the climate change response. If even a very basic conception of a right to health cannot be said to be protected under the current climate change response, then this will present a strong case for claiming that the climate change response is unjust and unable to meet even the most minimal demands of justice. For all of the reasons above, a right to health will therefore be defined as minimally as possible.

It is important to base the definition on an existing one, so that it is recognisable to policymakers. As was discussed above, it is imperative for policymakers to adapt climate-change-related rights into their discussions on climate change, because this will humanise climate change and may add a sense of urgency to the climate change response. The right to health will therefore be defined in line with Article Twenty-Five of the UNDHR as 'a right to standard of living adequate for health'. Adequate will here be taken to mean sustaining life at a minimally decent level. Minimally decent is a concept that has been defined in many different ways, but the capability approach taken by Martha Nussbaum provides a particularly useful definition because of its relation to health. Nussbaum (2004: 13) explains that defining a minimally decent life requires thinking about the prerequisites for living a life that is 'fully human rather than subhuman, a life worthy of the dignity of the human being'. In order to consider what this requires, Nussbaum takes a capability approach, and therefore argues that living such a life is a matter of what people are able to do, or in other words the kind of life they are able to lead. In this sense, a minimally decent life is contingent on several 'central human capabilities', including health, defined by Nussbaum (2006: 23) as 'being able to have good health; to be adequately nourished; to have adequate shelter'.

This conception of health is useful because it captures the key risks of climate change defined above – food security, water shortages, displacement and health – in a very straightforward manner. Adequate nourishment and adequate shelter capture the importance of food, water and displacement, all of which threaten human health. In this sense, health is defined to reflect the realities of the climate change problem and the threats it presents to humans. This is important in terms of defining a moral threshold that can be used for the evaluation of current practice, because it also allows the book to set out a precisely defined moral threshold that accurately reflects why climate change presents a threat to humans. Furthermore, the definition is very straightforward and understandable for policymakers because it is based on terms whose importance they are likely to understand (health, nourishment and shelter). Finally, Nussbaum's approach has been defended and used by cosmopolitans for years, which makes it recognisable to fellow scholars who are interested seeing how a climate justice approach can be used to assess current practice. Speaking to these scholars is highly important, since the book hopes to convince them of the value of such an assessment. For all of these reasons, the book will adapt Nussbaum's approach and define the right to health as 'the right to a standard of living adequate for health, including the right to be adequately nourished and have adequate shelter'. The chapter now turns to outlining how this right can be protected by multilateral and transnational climate change responses.

HOW CAN THE HUMAN RIGHT TO HEALTH BE PROTECTED?

The latest IPCC report explains that in order to avoid dangerous climate change, global temperature change must be kept at or below 2°C compared to pre-industrial levels (IPCC 2014a: 14). Importantly, there is a relationship between 'the right to a standard of living adequate for health, including the right to be adequately nourished and have adequate shelter' and the 2°C threshold. A 2°C rise is projected to cause major ice melting, wildfires, ocean acidification and heatwaves (Hansen et al. 2013: 3). These effects of climate change will result in loss of life, forced migration, increased stress on water and food resources and the spread of disease, all of which threaten human health (Hansen et al. 2013: 3). For this reason, world health experts have concluded with 'very high confidence' that climate change will contribute to the global burden of disease and premature death, and that a rise of 2°C will have deleterious consequences for human health (Hansen et al. 2013: 8).

This strongly suggests that protecting 'the right to a standard of living adequate for health, including the right to be adequately nourished and have adequate shelter' will require, at the very least, that global temperatures are kept at or below 2°C relative to pre-industrial levels. As was explained above, living an adequate life is a matter of what people are able to do, or in other words, the kind of lives they are able to lead. Crossing the 2°C threshold will threaten the central capabilities required for adequate health. The picture of a 2°C warming painted by climate scientists is a life that is consumed with the struggle to survive. Access to water, food and shelter will all be compromised, leaving those who are affected in a precarious position where their very survival is constantly threatened. Diseases will spread, and people will be displaced and unable to live in adequate shelter that provides proper sanitation or access to sufficient nutrition. A standard of living adequate for health does not seem possible under these conditions. Chapter 3 will expand on this idea and discuss what exactly must be done, in terms of mitigating emissions, to avoid crossing the 2°C threshold.

Of course, fully protecting 'the right to a standard of living adequate for health, including the right to be adequately nourished and have adequate shelter' will require more than simply mitigating emissions. Climate change may produce ill effects well before the 2°C limit is reached (Jamison et al. 2006), and scientists are confident that mitigation will not eliminate all climate change impacts (IPCC 2007: 73). Adaptation will therefore be a crucial part of protecting the right to health. Areas with weak health infrastructure – mostly in developing countries – will be the least able to cope with health problems (WHO 2017). They may therefore require assistance, especially given that many of the adaptive strategies that protect human health will be very expensive: warning systems, air conditioning, pollution controls, housing, storm shelters, vector control, vaccination, water treatment and sanitation are all costly measures (WHO 2003: 226). For this reason, finance for adaptation, as well as sharing of technologies and training, will all be a necessary part of protecting the human right to health. Adaptation is key in the case of health because strengthening health systems could significantly reduce the burden of disease and promote population health (WHO 2013: 1). Chapter 3 will therefore focus not only avoiding crossing the 2°C threshold, but also what will be required in terms of adaptation.

Finally, even though the right to health has been defined as minimal, it is clear that more must be done. The right to health is linked to other issues of global justice. A useful way to illustrate this point is to outline the social determinants of health, which encompass the ways in which health is shaped by various social factors and living conditions (Labonte 2014: 1). Some examples of social determinants of health are income and employment, education, health systems, social protection, the built environment and social patterns of exclusion (Labonte 2014: 8). Most of these social determinants of health overlap and affect one another. For example, two of the most impactful social determinants of health are income and education, and these factors seem to be intertwined in a number of ways. Studies show that the most disadvantaged members of society, especially those with below poverty-level incomes or without a high school diploma, generally experience the worst health, and even those with intermediate income appear less healthy than the most affluent and educated members of society (Labonte 2014: 9). The direct impact of income is related to having more economic resources and thus having access to healthier nutrition, housing or neighbourhood conditions, or less stress due to the availability of more resources to cope with daily challenges (Labonte 2014: 9). Income is linked to education, because poor families may not be able to send their children to school or university, and without degrees, these children will not be able to get high-paying jobs in their adult life and may not send their own children to school – meaning the cycle of lack of education and poverty continues, which negatively impacts health.

Some of the above social determinants are linked to climate change, since climate change affects the built environment and health systems, but many are also linked to overarching global inequality issues such as unequal income, education and unemployment, which are important to highlight as a cosmopolitan. Cosmopolitan justice is concerned with the fair treatment of morally equal human beings, and the social determinants of health are clearly not distributed in an equal manner across the globe, or even within nations. If they were, then every individual on the planet would be in good health. The high rates of childhood mortality and low life expectancy in poorer countries, and the high levels of health inequality even within the richest countries, suggest that this is not the case. For this reason, protecting the right to health adequately will include taking aspects of the social determinants of health seriously. However, for the purposes of this book, which seeks to assess current action on climate change from a global justice perspective, the failure to protect a minimally conceived right to health will provide very strong evidence that the current climate action does not represent a just response to climate change. The above serves only to acknowledge that health is a complex issue, which is being simplified for the sake of this book in order to demonstrate that climate change action cannot be said to represent even the most minimal protection of the right to health. For this reason, the climate justice position will be grounded in the right to health, defined in a very minimal sense as 'the right to a standard of living adequate for health, including the right to be adequately nourished and have adequate shelter'.

CONCLUSION

The first part of this book focuses on four normative questions in order to develop a climate justice account: who will be most affected, what exactly is at stake, what action must be taken and who should be responsible for this action. The previous chapter focused on the first question, defining the scope of justice as both relational and non-relational. This second chapter answered the second question by setting out the grounds of climate justice: the human right to health, defined as 'the right to a standard of living adequate for health, including the right to be adequately nourished and have adequate shelter'. The chapter first set out reasons for using a human rights approach, explaining that such an approach is commonly used to define moral thresholds, is recognised as important by policymakers at the global level, 'humanises' the climate change problem, adds urgency to the debate and is an increasingly popular approach among climate justice scholars. The chapter then defended the prioritisation of health over other human rights, explaining that the human right to health provides a very strong foundation for climate justice because it captures the threats climate change poses to humans more comprehensively than other key human rights. Finally, the chapter provided a definition of the right to health, based on Nussbaum's model for an adequate life, and began to explain what action is necessary in order to protect this right. This will be expanded on in Chapter 3, which concerns answering the final two questions: what action must be taken in the face of climate change, and who should be responsible for this action.

NOTE

1. Alasdair Cochrane (2012: 659) argues that certain animals have the same basic rights as humans, because 'it is surely only sensible to recognise that all sentient creatures have at least some very basic interests which are sufficiently strong, all things considered, to ground some duties on the part of others'. For example, many higher animals have appetites and can feel the pain of ill health, extreme thirst and starvation, which suggests they may have a right to be protected from these harms. Cochrane's arguments are very persuasive, and intuitively it seems that animal rights should be part of the climate justice account defended here. However, this book is based in a cosmopolitan approach, and therefore takes individual humans and their moral worth as the ultimate unit for concern. Although there could of course be a cosmopolitan framework which includes animals, this area of cosmopolitanism is somewhat underdeveloped. There are a number of scholars working on the subject (Cochrane 2013; Horta 2013; Pepper 2017), but, as it stands, it would be an arduous and time-consuming task to apply this emerging framework to the climate change problem. A comprehensive animal-focused climate justice framework has so far not been developed, and attempting to develop one here is beyond the scope of this book.

Chapter 3

THE DEMANDS OF CLIMATE JUSTICE

INTRODUCTION

Part I of this book develops a climate justice account that clarifies what exactly is normatively expected of climate change actors. Once this is clear, the book can turn to assessing to what extent policymakers have met these normative expectations in Part II. So far, Chapter 1 has set out a scope of justice that is both relational and non-relational in order to include the primary victims of climate change in the realm of moral concern. Chapter 2 has set out a minimum threshold that must be protected by policymakers: the human right to health, defined as 'the right to a standard of living adequate for health, including the right to be adequately nourished and have adequate shelter'. The current chapter completes Part I by focusing on what action must be taken in the face of climate change and who should be responsible this action. In answering these two questions, the chapter will set out three demands of climate justice that must underwrite a more just global response to climate change. Setting out these demands enables the book to move on to assessing to what extent the global response to climate change has met these demands in Part II.

The chapter is organised into three main parts, each one developing a demand of justice. The first part focuses on what action must be taken on climate change. Here, the chapter puts forward that the human right to health must be protected: temperatures must be kept below the 2°C threshold, and adaptation must be prioritised alongside mitigation. The second and third parts of the chapter then turn to the question of who should be responsible for climate change action. Part two considers the relationship between developed and less developed countries, focusing on what this relationship means for the fair allocation of responsibilities. In doing so, the chapter develops the 'Polluter's Ability to Pay' or PATP model, which holds states accountable for climate action in line with their emissions and wealth levels. The third part of the chapter then explores whether there are any other actors who should be held to account for climate change action. Focusing on the relationship between those who cause climate change and those who suffer from its effects, the chapter puts forward that non-state actors are harming climate change victims by perpetuating a wider system of injustice that threatens their right to health and should therefore be held accountable for climate change action. The chapter concludes with a summary of the arguments made in Part I of the book, as well as an outline of what is to come in Part II.

PROTECTING THE HUMAN RIGHT TO HEALTH

The book has so far explained that the primary victims of climate change will be future generations and those living in less developed countries, and that both must be included in the realm of moral concern, or scope of climate justice. Furthermore, the book has set out what is at stake for these two groups of victims: the human right to health, defined as 'the right to a standard of living adequate for health, including the right to be adequately nourished and have adequate shelter'. Protecting this right will require policymakers to, at the very minimum, keep global temperature changes to 2°C or less, as was briefly discussed in the previous chapter. Although climate change effects are already occurring and will increasingly affect present generations, setting a 2°C minimum necessarily implies considering what is owed to future generations, because a temperature change of 2°C is not predicted to occur until 2050–2100 (IPCC 2014a: 14). The first part of this chapter will therefore focus on why future generations are owed the protection of their human right to health under the non-relational scope set out in this book.

Of course, protecting the right to health will also require adaptation measures, as was briefly discussed in the previous chapter. This is especially true for less developed countries. It is widely accepted that wealthy nations have a greater capacity to adapt than poorer ones because they have the economic resources to invest in adaptive measures and to bear the costs of adaptation (WHO 2003: 226). By contrast, the feasibility of adaptation options for many poor countries is constrained by a lack of resources (WHO 2003: 226). This may imply that developed countries have a moral obligation to help developed countries with adaptation costs. These moral obligations are important to consider as part of the climate justice account, because they will no doubt affect to what extent the human right to health can be protected in less developed countries. The first part of this chapter will therefore discuss mitigation and adaptation in turn. This discussion will allow the chapter to set out the first demand of climate justice, which centres around defining exactly what action must be taken on climate change.

What is Owed to Future Generations?

The scope of justice defended in this book is based, in the first instance, on a non-relational moral minimum: the human right to health. The book therefore assumes that present and future generations must be considered as morally equal, and that both generations deserve to have their right to health protected. As was explained in Chapter 1, under a non-relational scope, time of birth is – like place of birth – a matter of luck, rendering time morally arbitrary. This means all humans are entitled to the same rights, no matter when they are born, simply because they are human beings. More specifically, according to the scope and grounds of climate justice set out in this book, all humans, unconditionally, have 'the right to a standard of living adequate for health, including the right to be adequately nourished and have adequate shelter', no matter when they are born. In this sense, what is owed to future generations is relatively straightforward: it is the protection of their human right to health.

Although present generations do not violate this right at the present moment, because future people do not yet exist, their actions are assured to do so in the future. This time-lag between actions today and the violation of rights in the future is morally irrelevant according to a non-relational scope of justice. Humans have the right to health, no matter the time of their birth. It is therefore morally wrong to act in ways that cause an individual's right to health to be violated, whether the right is violated immediately or at some point in the future. This has important policy implications: if we can reasonably predict that the right to health will be threatened by policy decisions made today, then this must inform these decisions (Vanderheiden 2008: 132). Steven Vanderheiden (2008: 136) refers to this as the rational capacity for foresight: 'the fact that we can foresee having obligations has moral consequences for us'. Considering the vast amount of evidence on climate change and its effects on health, outlined in Chapter 2, policymakers can be all but certain that the right to health of future generations will be threatened by climate change. If this evidence is accepted as credible, there indeed exists 'rational foresight' that the right to health of future generations is threatened by climate change. Policymakers are therefore under a moral obligation to protect this right.

However, some scholars argue that protection of the human rights of future persons is not morally necessary. The most well-known argument against protecting the rights of future generations is presented by Derek Parfit. Parfit sets up a puzzle he refers to as the 'non-identity problem', which asserts that the harm committed against future generations is not morally problematic as long as these individuals have a life worth living. The non-identity problem rests on the idea that the exact moment an individual is conceived is highly important: if an individual had been conceived 'even an hour after their actual conception', they would not be the same individual, but someone else, because the genetic makeup of the individual would be based on a different sperm (Parfit 2010: 113). This is relevant to the climate change problem for the following reason. When current generations choose to pollute, this means that there will be a higher standard of living for the next few centuries, resulting from the profits of energy consumption. According to the logic underpinning the non-identity problem, this will directly affect conception times, because being richer implies that people will marry different people over time, and even in the same marriages, the children will be conceived at different times (Parfit 2010: 113). This implies that current generations can directly affect which individuals are conceived by choosing not to lower greenhouse gas (GHG) emissions, which will affect wealth levels, and ultimately conception times.

According to the non-identity problem, the power of current generations to affect conception times implies that future generations cannot be harmed, because they are benefited overall by being born in the first place (Parfit 2010: 114). In this sense, even if high GHG emissions levels negatively affect the environment, existing and suffering from the effects of climate change is better than never having existed at all. Therefore, the decision of present generations to emit GHGs, which leads to the birth of a specific future generation, does not harm that specific generation. If GHG emissions levels had been lowered, then specific future generations would never have been born. Overall, then, there is no moral wrong in emitting GHGs, because this leads to future generations benefiting by being born in the first place (Parfit 2010: 116). However, there is an important caveat to the non-identity problem, which is that individuals must be able to lead a 'life worth living' (Parfit 2010: 113). According to the non-identity problem, if polluting results in lives which are worth living, then 'we know that our choice [to pollute] will not be worse for [future generations]' (Parfit 2010: 116). This is an important caveat, because it is questionable whether future generations will be able to lead such a life if their right to health is violated.

It was explained in the previous chapter that a minimally decent life, or 'life worth living' à la Parfit, requires thinking about the prerequisites for living a life that is 'fully human rather than subhuman, a life worthy of the dignity of the human being' (Nussbaum 2004: 13). Chapter 2 argued that living such a life is a matter of what people are able to do, or in other words the kind of life they are able to lead. In the case of health, humans must 'be able to have good health; to be adequately nourished; to have adequate shelter' (Nussbaum 2006: 23). Climate change threatens all three of these abilities very severely, as was discussed in Chapter 2. Climate change is predicted to lead to an increase in diseases such as malaria, diarrheal diseases, infectious disease such as HIV and AIDS as well as an increase in serious cardio-respiratory problems, all of which threaten the ability to have good health (IPCC 2014a: 11). The ability to be adequately nourished will also be affected by climate change, which is expected to pose risks to food security and threaten water resources, leading to an increase in life-threatening dehydration and malnourishment (IPCC 2014a: 13). Finally, if crops fail, water supplies run short, sea levels rise and major weather events result in vast areas becoming inhabitable, the ability to have secure shelter will be an impossibility for millions (Biermann and Boas 2010: 68).

If the central capabilities required for adequate health are all threatened by climate change, it is hard to imagine how future peoples will lead a minimally decent life. The picture painted by climate scientists is of a life consumed with the struggle to survive. Access to water, food and shelter will all be compromised, leaving those who are affected in a precarious position where their very survival is constantly threatened. A standard of living adequate for health does not seem possible under these conditions. This suggests that the non-identity problem falsely assumes that future generations will be able to lead a life worth living, or 'fully human rather than subhuman', even if pollution occurs. In this sense, the non-identity problem does not seem to present a convincing account against the rights of future generations, because it does not fully take the realities of climate change into account.

Furthermore, even if sceptics believe that future humans will be able to live a life worth living despite climate change, the threat to their human right to health is an important moral wrong that is not fully acknowledged by the non-identity problem. Instead, the non-identity problem purports that future generations are overall benefited from being born and can therefore not be harmed by current pollution. This seems counter-intuitive, because even if an action that violates a person's human rights benefits this person in some way, this does not cancel out or detract from the moral wrong of the rights violation. For example, even if an individual who was held in a Nazi concentration camp feels that this experience bestowed her with a deeper appreciation of life than if she had not been a prisoner, this does not negate the fact that her human rights were violated (Woodward 1986: 809). In this sense, even if high-emitting present generations benefit future generations by ensuring that they are born, which means they are better off than if they had never been born at all, this does not detract from the moral wrong which is committed against future individuals whose right to health is violated. The non-identity problem does not address this moral harm, and for this reason it does not provide an adequate defence against the rights of future generations.

A different, and less convincing, argument against protecting the rights of future generations is put forward by'pragmatists', who claim that future generations will be better off than present generations and will therefore have the capacity to address climate change (see for example Lomborg 2001; Collier 2010). The problem with this type of argument is that it is not certain future generations will be wealthier, because the level of cost to tackle climate change may increase at a greater speed than the level of wealth (Caney 2009: 172). In addition, the pragmatic argument overlooks that it may be cheaper to tackle the problem earlier rather than later, or even that there may be a 'tipping point' of irreversible damage (Caney 2009: 174). Failing to curb emissions will lead to increasingly dangerous climate change. Eventually, inaction will lead to irreversible damages which are incompatible with human life. These irreversible changes include mass extinction of animals, changes in marine ecosystem productivity, damage to fisheries, changes in oceanic oxygen concentrations and decreased terrestrial vegetation (IPCC 2007: 56). For this reason, claiming that future generations will be better off and therefore better equipped to deal with the consequences of climate change presents a fundamental misunderstanding of the best available scientific evidence. The argument that future generations will be better off is therefore dismissed at this point. It is clear that according to the non-relational scope and grounds set out in this book, future generations have a right to health in the same sense that present generations do, and this right must be protected.

What Is Required for the Protection of the Human Right to Health?

Substantial action on climate change will be required in order to protect the right to health of future generations. Chapter 2 explained that, at minimum, emissions must be kept in check enough to keep global temperature changes at or below 2°C compared to pre-industrial times. Surpassing this threshold will lead to 'dangerous' climate change: wide-scale floods, droughts, heatwaves, sea-level rises and forced migration, all of which threaten the right to health. For this reason, world health experts have concluded with 'very high confidence' that climate change will contribute to the global burden of disease and premature death, and that a rise of 2°C will have deleterious consequences for human health (Hansen et al. 2013: 8). Taking these considerations into account, the stance on the protection of the right to health of future generations is the following:

1.   a) Global temperature changes must be kept at or below 2°C.

This is the first explicit demand of justice that must be met in order to achieve a condition of climate justice according to the account defended in this book. Keeping global temperatures below 2°C will require substantial emissions reductions. The Intergovernmental Panel on Climate Change (IPCC) claims, at the time of writing, that GHG emissions would have to be cut by 40–70% by 2050 compared to 2010 and would need to be near zero or below in 2100 (IPCC 2014a: 8). The latest IPCC report explains that GHG emissions have continued to increase over 1970 to 2010 despite a growing number of climate change mitigation policies (IPCC 2014a: 5). Indeed, the growth rate of emissions increased from 1.5% a year in the period 1980–2000 to 3% a year in 2000–12 (IPCC 2014a: 1). Global emissions have only recently begun to slow down, increasing by 0.7% in 2014, stalling in 2015 and growing by 0.2% in 2016 (Carbon Brief 2016). Worryingly, it seems that this brief plateau in emissions is unlikely to last, with emissions rising by 2% in 2017 (Carbon Brief 2017). This indicates that even a very minimally conceived right to health requires urgent change in behaviour at the global level to keep global temperature change at or below 2°C. Therefore, it seems that in order to prevent the violation of the right to health, actors around the world will have to lower emissions substantially. The second part of this book will assess to what extent multilateral and transnational actors are working towards these emissions reductions. Of course, the 2°C limit is only one part of the story. Protecting the right to health will require more than simply keeping emissions at bay – adaptation will play a key part in its protection.

The Importance of Adaptation

Adaptation is key for protecting human health because strengthening health systems could significantly reduce the burden of disease and promote population health (WHO 2013: 1). For example, the threat of changing patterns of disease due to insect-borne infections can be countered through adaptation measures such as vector control, promotion of mosquito nets, new vaccines or rapid and effective diagnosis and treatment (Costello et al. 2009: 1702). In addition, early warning systems can prevent many cases of ill health, especially in the context of heatstroke, extreme weather events and disease outbreaks (Costello et al. 2009: 1710). Furthermore, when combating water shortages, systems for safely storing and treating water and technologies for using alternative supplies of water, such as waste-water recycling and desalination, could make a significant difference to health outcomes (Costello et al. 2009: 373). Finally, the design of houses and settlements could protect health through protection against thermal extremes, disaster-proofing, barriers and deterrents to infectious disease vectors, and energy efficiency (Costello et al. 1717). However, many of the adaptive strategies that protect human health will be very expensive: warning systems, air conditioning, pollution controls, housing, storm shelters, vector control, vaccination, water treatment and sanitation are all costly measures (WHO 2003: 226). It is not just the expense of adaptation that poses a problem. Cognitive, behavioural, political, social, institutional and cultural constraints limit both the implementation and effectiveness of adaptation measures (IPCC 2007: 56). From this, it seems obvious that some countries, in particular less developed countries, may require assistance from richer states or non-state actors in order to adapt to climate change effectively, thereby protecting the human right to health. Considering that those living in less developed countries are predicted to be the primary victims of climate change, the importance of adaptation cannot be ignored.

In fact, health adaptation is increasingly emerging as a concern among policymakers at the local, regional and global level (Bowen and Friel 2012). Given the importance of adaptation, it seems that Demand One, as defined above, is somewhat lacking – protecting the right to health will require more than simply keeping emissions at bay. Adaptation is clearly key when it comes to protecting those living in less developed countries who are alive today, and who are yet to be born. Adaptation will entail a number of different measures, rather than one sum of money, or one particular transfer of technology. Defining a demand around adaptation is therefore difficult without oversimplifying the complexity of adaptation needs. For this reason, the book will set out a broad demand: that adaptation is prioritised alongside mitigation, as an equally important part of the climate change response. Although this might seem like too low a threshold, it is important to examine whether even this very minimal demand is being met by the climate change response. If it is not, then this will be very revealing of the lack of action that has been taken on adaptation so far. Demand One will therefore be expanded to include a second part:

1. a) Global temperature changes must be kept at or below 2°C.
    b) Adaptation must be prioritised alongside mitigation.

Now that the chapter has explained what protecting the right to health requires, it turns to discussing who is responsible for this protection, both in terms of lowering emissions and supporting adaptation. The chapter will first discuss the responsibilities of states, before considering whether non-state actors should be held to account for climate change action alongside states.

THE RESPONSIBILITIES OF STATES

Assessing how to allocate responsibilities for climate change is impossible without considering the relationship between developed and less developed countries. The impacts of climate change will fall disproportionately upon less developed countries. And yet, developed countries are the main cause of climate change. These realities of the climate change problem create a relationship between developed and less developed countries, because developed countries are engaging in behaviour that endangers the people living in less developed countries. This relationship, in turn, raises profound questions of fairness that have hampered climate change action for decades. Less developed countries find it unfair that they must suffer from the consequences of climate change when they have not caused this problem. These countries therefore often argue that they deserve compensation, and that developed countries should act first (Harris 2010: 91). Less developed countries also find it unfair that they must curb emissions when developed countries had the chance to emit freely (Harris 2010: 91). The argument here is that that they need more time before they lower emissions, so that they can become wealthy enough to develop green technology and contribute to the global costs of climate change mitigation and adaptation.

At the same time, it is clear that if less developed countries pursue their own economic development with the same disregard for the natural environment that developed countries displayed, it will dramatically worsen the predicted effects of climate change (Shue 1999: 531). This is especially true considering that less developed countries are projected to contribute 45% of global GHG emissions by 2050 (Lawrence 2014: 13). Developed countries are therefore keen to see less developed countries contribute to climate change action. In fact, developed countries have been known to be hesitant to act without the commitment of less developed countries, which is one of the main reasons the USA did not ratify the Kyoto Protocol, and why Donald Trump decided to pull the USA out of the Paris Agreement. Developed countries do not think that it is fair to place climate change responsibility solely on richer countries. They want to see key less developed countries, including China and India, do their fair share. These different perceptions of what is fair make it extremely difficult to allocate responsibility for adaptation and mitigation. Perceptions of fairness have made it near impossible for states to agree on climate action.

What is needed is relational analysis that takes into account issues of fairness stemming from the relationship between developed and less developed countries. The chapter will therefore now apply the relational side of the climate justice account to conduct an analysis of the relationship between developed and less developed countries. Conducting relational analysis will allow the chapter to put forward the book's second demand of climate justice, which centres around who should be responsible for climate change action.

The Categorisation of Countries into Developed and Less Developed

Before the relationship between developed and less developed countries can be analysed, it is important to stress that 'less developed country' is not a straightforward or unproblematic categorisation. This can be seen in the many ways the category is defined. The World Bank, for example, defines countries as less developed according to a simple economic calculation. Countries with a gross national income of $12,476 USD or less are defined as developing countries until December 2017 (World Bank 2017). The United Nations Development Programme (UNDP) takes a more complex approach, dubbed the 'Human Development Index'. This index was created to emphasise that people and their capabilities should be the ultimate criteria for assessing the development of a country, rather than economic growth alone. The Human Development Index takes into account three factors: life expectancy, level of education and standard of living. The scores of these factors are then aggregated into a composite index using geometric mean.

The difference in the approaches of the UNDP and the World Bank illustrates that there is significant disagreement over which factors to take into account. For this reason, there is no one standard definition of 'less developed'. To add to the complexity, there is a category which captures five major emerging economies referred to as BRICS (Brazil, Russia, India, China and South Africa). Although all of these countries, except Russia, are considered less developed according to the World Bank, the BRICS countries are distinguished from other less developed countries by their large, fast-growing economies and significant influence on regional and global affairs (all five are G20 members, for example). Furthermore, the BRICS countries represent 26% of the planet's land mass, are home to 46% of the world's population and account for 18% of the world's GDP. The economic and political power of these countries illustrates that the category of 'less developed countries' captures countries that are incredibly diverse – China and Somalia, for example, are both considered less developed according to the World Bank.

And yet, the distinction between less developed and developed countries is especially important in the case of climate change. Under the Kyoto Protocol, in force until 2020, only certain 'developed' countries are held to direct account for financial contributions and lowering of emissions. This will be further discussed in Chapter 5, but it is worth briefly explaining the categories employed by the United Nations Framework Convention on Climate Change (UNFCCC) here. The UNFCCC currently has four categories of countries: Annex I, Annex II, non-Annex I and Least Developed Countries. Annex I countries are 'industrialised' and held to account for emissions reductions, and the Annex II countries, which represent the OECD members of Annex I, are held to account for financial contributions under the Kyoto Protocol. Non-Annex I countries, a category which includes 148 countries including Brazil, China, India and South Africa, are considered 'mostly developing countries' and are not held to account for emissions reductions or financial contributions under the Kyoto Protocol. Finally, out of the 148 Non-Annex I countries, there are forty-nine 'Least Developed Countries', including Angola, Rwanda, Uganda and Somalia, which are considered especially vulnerable to climate change because of their limited capacity to respond and adapt to its adverse effects.

Because this book concerns itself with climate change, the categories of the UNFCCC will be used to define which countries are less developed (non-Annex I) and which countries are particularly vulnerable to climate change (Least Developed Countries) in the remainder of the book. However, it is acknowledged here that this is not the only way to categorise these countries, and that the book is merely following one categorisation which is well established, and more importantly, relevant to the climate change problem. The discussion below, and in the remainder of the book, will aim to take the complexities of categorising countries into account when discussing the relationship between less developed and developed countries.

The Importance of Fairness

The relational analysis of developed and less developed countries, or more specifically what this relationship implies for climate responsibilities, will here centre on fairness. Fairness is extremely important in the case of climate change because an allocation of responsibilities that is perceived as unfair by either developed or less developed countries is likely to be rejected. If responsibilities are rejected, climate change action will be drastically undermined due to lack of participation. Considering the urgency of the climate change problem, and the non-relational right to health at stake, action on climate change is morally required. If the issue of fairness is important to ensure this action is taken, then finding a fair solution is, by this logic, morally important. For this reason, it is crucial to consider what both developed and less developed countries perceive to be a fair distribution of responsibilities in the case of climate change. This focus on fair distribution will inform the relational analysis and enable the chapter to define a demand of justice that centres around which states should be responsible for how much climate change action.

Fair distribution will here be defined in line with Brian Barry's definition: 'people can accept not merely in the sense that they cannot reasonably expect to get more, but in the stronger sense that they cannot reasonably claim more, morally speaking, as they can reasonably accept the distribution and have no moral claim for a different distribution' (Barry 1989: 8). Barry's argument rests on the idea that if a party can reasonably agree to a model of distribution, the party accepts distribution as fair. This definition of fairness has been well established in political philosophy. Although there exist other conceptions of fairness, Barry's conception is useful in the sense that it provides a moral argument for exploring the relationship between parties in order to understand what these parties could reasonably agree to. Barry provides a manner in which to explore the realities of existing relationships and define what is morally fair in these relationships. For this reason, the book will adapt Barry's conception of moral fairness, in order to discuss how to fairly distribute benefits and burdens between developed and less developed countries.

Although the book makes use of Barry's position, his position will necessarily have to be adapted in order to ensure that it is compatible with the climate justice account defended here. While the demand of justice developed below is a normative one, and reflects what countries ought to do according to the climate justice account, the book will base this demand on a discussion of what countries actually find fair, and would therefore agree to. In other words, the chapter explores the realities of the relationship between developed and less developed countries in order to consider what justice demands in the case of this relationship. This differs from Barry's conception, which is moral rather than empirical in the sense that he is not concerned with what parties want, but rather solely concerned with what parties ought to do.

The book adds an empirical element to the discussion in order to apply the mixed scope of justice defended in Chapter 1. The scope of climate justice employed in this book is in part relational, which enables the book to explore the realities of global relationships caused by climate change in order to explicate demands of justice. In this sense, the book aims to explore the realities of what countries find fair in order to explicate demands of justice which are specific to the relationship between developed and less developed countries. If the current chapter merely assigned moral duties to countries without considering the realities of what these countries find fair, then the book would not be conducting a relational analysis. For this reason, although the book makes use of Barry's position, the position is adapted in the sense that the realities of what countries find fair is included. In this way, Barry's notion of fairness is used to provide a blueprint for a normative relational discussion, which results in the formulation of a demand of climate justice. This demand is not based on a purely moral discussion, but on an empirical discussion of the reality of the relationship between developed and less developed countries that informs this moral discussion. The book now turns to assessing what these countries perceive as fair and how this relates to existing cosmopolitan models of climate change burden distribution. This will allow the chapter to develop its own model of distribution, referred to as the 'Polluter's Ability to Pay' or PATP, for short.

Existing Perceptions of Fairness

Let us begin by considering what less developed countries find fair. According to Vanderheiden (2008), less developed countries have three main concerns related to the issue of fairness. The first relates to the idea that developed countries have contributed most to the climate change problem and should therefore pay more towards the cost of combating its effects (Vanderheiden 2008: 67). This implies that less developed countries will perceive a distribution of responsibility as unfair if these are not in line with historical contribution. Second, less developed countries believe that they face greater immediate problems that must be addressed before they can agree to help with climate change action (Vanderheiden 2008: 67). It is difficult to see why less developed countries should divert their attention from their own worst problems in order to help with problems that, for them, are far less immediate and deadly (Shue 1999: 543). This implies that a fair allocation of responsibilities should reflect the importance of addressing urgent basic needs, because otherwise less developed countries cannot reasonably agree to it. Third, less developed countries believe that they should have a right to develop and should therefore not have to cut their emissions as drastically as developed countries until such development is achieved (Vanderheiden 2008: 67). This is a complex concern, because less developed countries should not have to accept a climate change deal where they are essentially blocked from developing, but they cannot be left out of the climate deal entirely, because universal participation is necessary (Vanderheiden 2008: 67).

To make matters more complicated, developed countries will see little point in acting if the less developed countries are holding back (Harris 2010: 90). In this sense, the primary concern for developed countries is that they are not pursuing climate change action on their own. The perceived unfairness of less developed countries not doing their fair share is the primary reason that US president Bush did not ratify the Kyoto Protocol, and why Donald Trump pulled the USA out of the Paris Agreement. In sum, the three main concerns of the less developed countries are acceptance of responsibility by the rich, ensuring basic needs are met and guaranteeing that there is an opportunity to develop. The main concern of developed countries, conversely, is that they are not alone in acting on climate change. If these are the main concerns in terms of fairness, all four concerns must be addressed to make the allocation of responsibility reasonably acceptable to all parties.

Of course, it is important to remember that not all three of the less developed country concerns will be equally important to every country in this vast category. For example, Least Developed Countries will be more concerned about addressing urgent needs than having a right to develop, because they undoubtedly have greater urgent needs to address than BRICS countries. Richer less developed countries like China and Brazil, on the other hand, may be more concerned with having a right to develop, because these countries are well on their way to becoming developed countries, at least according to the BRICS categorisation, and will feel strongly about not having this development impinged. Keeping these complexities in mind, the chapter now turns to discussing how these existing perceptions of fairness match up to cosmopolitan models of responsibility.

Existing Models for Allocating Climate Change Responsibilities

There are three well-established cosmopolitan positions on climate responsibility: Polluter Pays, Ability to Pay or a mixture of both. These existing models have been at the centre of climate justice literature for decades. The book aims to engage with this literature and will therefore now assess whether these models are in line with what developed and less developed countries perceive to be fair. Through this assessment, the chapter will develop a new model of responsibility that improves on existing accounts – the 'Polluter's Ability to Pay' or PATP model. Let us begin by considering that the Polluter Pays Principle, commonly referred to as the PPP, is based on examining who caused the problem, and using this information to determine who should pay (and how much) for climate change action. The PPP is based on a well-known conception of fairness whereby we 'clean up our own mess' (Shue 1999: 533). According to advocates of the PPP, countries that have emitted the most, and/or continue to emit the most, are responsible for paying for most of the damages caused by these emissions (including adaptation costs). Proponents of the PPP usually conclude that developed countries should bear most of the burdens of climate change due to their high GHG emissions (Gardiner 2004: 579). In this way, the PPP is based entirely on taking responsibility for high GHG emissions by paying to fix the problems associated with these emissions.

The PPP has an important advantage: it has been strongly defended by China, Brazil and other less developed countries in climate change negotiations (Page 2012: 305), implying that the less developed countries perceive it to be fair. This is unsurprising, because the PPP seems to fit in with the three concerns of less developed countries outlined above. The PPP explicitly places responsibility on developed countries, allows less developed countries to address urgent needs since they are not required to act on climate change, and also allows these countries to develop without being burdened by costs, since they are not expected to pay for emissions reductions unless their emissions reach a certain level. In addition, this model could be perceived to be fair by developed countries, who would not have to 'go it alone' because climate responsibilities would be based on a clear emissions threshold and countries would have to accept responsibility once they meet this threshold. This may imply responsibilities for high-emitting BRICS countries, for example.

However, the PPP could be considered unfair by some high-emitting less developed countries, because the correlation between high emissions and wealth is not perfect. Requiring countries to pay in proportion to their emissions may perpetuate the poverty of some and reduce others to poverty. For example, China or India, countries which have high emissions, may not have the resources to pay for damages caused by these emissions because their level of wealth is not high enough to pay for damages (Caney 2010a: 212). This suggests that the PPP may result in an allocation of responsibility that cannot be reasonably agreed to by all parties. If high-emitting less developed countries are called on to pay for lowering emissions and this results in them not being able to address urgent needs of their population, or to continue their development because the cost of lowering emissions is too high, then it would be difficult for these countries to reasonably accept the distribution of benefits and burdens implied by the PPP.

In addition, the PPP does not seem to provide much incentive for cutting emissions. Continuing to emit, and therefore make profits to pay for climate change costs in line with PPP responsibilities, may be more cost-effective than having to pay for lowering emissions. In other words, some countries might want to 'pay their way' out of having to reduce emissions. This will surely be perceived as unfair to those countries who have invested in green technology, and will definitely be perceived as unfair by Least Developed Countries, who will be hardest hit by climate change impacts. For these Least Developed Countries, lowering of global emissions is a matter of life or death, having a place to live or being displaced, and being able to have good health or being threatened by disease. In this sense, Least Developed Countries may not find an allocation of responsibilities based solely on payments in line with emissions fair. For these reasons, the PPP does not represent a useful model for the allocation of climate responsibilities, because basing distribution on pollution levels alone may not be reasonably accepted by all. The chapter therefore turns to the next principle – the Ability to Pay.

The Ability to Pay Principle, commonly referred to as the ATP, asserts that the responsibility for addressing climate change should be borne by the wealthy, and, moreover, that this responsibility should increase in line with the agent's wealth (Caney 2010a: 213). The key difference between the PPP and the ATP is that the ATP is indifferent to who caused harm and focuses instead on who can rectify harm (Caney 2010a: 213). The ATP has several important advantages relating directly to the concerns of less developed countries outlined above. First, the responsibility to pay for climate change will fall mostly on the developed countries because they are the wealthiest countries, meaning they are held to account for their climate-change-causing actions. Secondly, if the responsibility to pay falls on a less developed country, this will only be when this country has the ability to pay. If a country is wealthy enough to pay for climate change action, or to reduce emissions, this seems to imply that other concerns, such as meeting the basic needs, may have already been met. This is particularly important in the case of Least Developed Countries, who are vulnerable to climate change and do not have the resources to address the problem. This addresses the second concern of the less developed countries, namely that urgent needs must be addressed before climate change can be acted upon. Finally, the fact that countries only have to pay when they are wealthy enough seems to imply that less developed countries would have time to develop to a certain point, at which they could begin to pay for climate change action. This addresses the third concern of the less developed countries (less developed countries have a right to develop), which is particularly important for richer less developed countries, who may feel that they have a right to develop to a certain level before being held accountable for climate change action. From the above, the ATP seems to be in line with what less developed countries consider fair, and what these countries could reasonably agree to.

However, the ATP suffers from a key disadvantage. Not taking levels of emissions into consideration implies that wealthy countries with low GHG emissions will be required to pay as much as wealthy countries with high GHG emissions. This could be perceived as unfair because wealthy countries which contribute to combating the climate change problem by limiting emissions are not rewarded for this behaviour, and are in fact treated in the same manner as high-emitting countries. This is an important weakness of the ATP because the book is concerned with a solution that can be reasonably accepted by all. It seems unreasonable for wealthy low-emitting nations, such as Iceland (average of 6.2 metric tons of carbon per capita per year), to accept that they have to pay at the same rate per emissions as rich polluting nations such as the USA (average of 17.6 metric tons of carbon per capita per year), with no recognition of their efforts besides lower payments. In addition, similarly to the PPP, the ATP does not provide an incentive for countries to lower emissions, because there is no reward for lowering emissions under this model. Instead, countries can simply 'pay their way', which seems counterproductive to the goals of lowering emissions worldwide. For these reasons, the ATP does not represent an adequate model for the allocation of climate responsibilities.

Considering the weaknesses of the ATP and PPP, perhaps a mixed approach is best able to represent a distribution of climate responsibility that can be reasonably accepted by less developed countries and developed countries alike. There is no one name for such a mixed approach, as each proponent has their own version of ATP/PPP combinations and they offer various names for these approaches. Risse (2008: 40), for example, suggests a 'Beneficiary Pays Principle', or BPP, that is based on an index that measures per-capita wealth and per-capita emission rates, and then groups countries into categories depending on their combined index, which weighs both criteria equally. The amount of emissions reduction or payments towards climate action for which a country is responsible is a function of this index (Risse 2008: 40). Under the BPP, a high number of countries would not incur any responsibility, because they are ranked too low according to either one or both criteria. To illustrate, countries which have high levels of pollution and high levels of wealth would be asked to reduce their pollution and to pay for global climate change costs, while countries of low wealth and high pollution would have to reduce their emissions as much as possible, and only when their wealth levels rise would they have to pay more towards climate change costs and further emissions reductions. Countries that have low emissions and low wealth would be excluded from action, and those with low emissions and high wealth would be asked to contribute financially but not lower their emissions.

This mixed approach avoids the PPP's problem of requiring high-polluting countries to lower emissions, even if these countries are not wealthy enough to be able to do so. Although some less developed countries have high emissions levels, they may not be able to afford to reduce emissions or pay for climate change costs as easily as other countries. This may interfere with their ability to address the basic needs of their populations or to pursue development, which are two of the less developed countries' concerns outlined above. At the same time, Risse's mixed approach acknowledges that some nations may be rich but have low emissions. As was discussed above, it is important to provide an incentive for countries with low emissions and to treat them differently than countries with high emissions. Risse's mixed approach has an advantage over the ATP in this regard, because the BPP highlights the fact that countries with low emissions and high emissions should be differentiated in terms of responsibilities. It may therefore be perceived as fairer than an approach based solely on wealth. The BPP also has an advantage over the PPP, because if countries are held to account in line with both emissions levels and wealth levels, it may not be cost effective to continue to emit and spend profits on climate change costs, since the richer a country is the more it will have to pay. Under a mixed model of responsibility it may be more cost-effective to pay for new technology and lose profits, because if per capita wealth goes down, this will be reflected in how much the country has to pay for climate change costs. In this sense, the BPP has the advantage of being fairer than either the ATP or PPP models on their own, which is important because it may be more easily accepted by both developed countries and less developed countries.

However, there is a second example of a mixed approach that has an important contribution to make in terms of addressing less developed countries' concerns. Simon Caney's (2010a: 215) ATP* model recognises a difference between (i) those whose wealth came about in ways which endangered the climate and (ii) those whose wealth came about in ways which did not endanger the climate. Caney (2010a: 215) believes that countries that fall into category (i) should be contributing more to climate change costs. This is in line with Risse's account, because Risse argues that high-polluting countries that are rich are responsible for both financial contribution and lowering emissions, whereas 'green' rich countries are only responsible for contributing financially. Interestingly, Caney (2010a: 218) qualifies his approach with the notion that a country should only bear the burden of climate change so long as doing so does not push that country beneath a decent standard of living. This argument is useful because it seems to be particularly sensitive to the second and third concerns of less developed countries, which are that less developed countries have urgent needs that have to be addressed before contributions to climate change are made, and that less developed countries should have the right to develop. If countries must only contribute so long as this does not push them beneath a decent standard of living, this implies that basic needs must be met and that countries must be developed enough to be able to contribute so that their contributions do not result in substandard living conditions for their population. This is especially fair to countries which face urgent needs, particularly those in the category of Least Developed, who would under Caney's model not be held to account for climate change action.

The mixed approaches defended by Risse and Caney have clear advantages over the ATP and PPP, especially in terms of taking the concerns of fairness into account. As was argued above, the allocation of responsibility is considered fair when all parties cannot reasonably reject a distribution. In order to meet this condition, there must be consideration of what less developed countries believe to be fair: the burden falling on developed states who are seen to have caused the problem, basic needs being met before any climate action is taken, and the right to develop being taken into account. However, it is also important that all states participate in climate action. Developed countries, especially those that are 'green', must find the model of responsibility fair, and should therefore not feel they are going it alone. The mixed approach is able to accommodate this concern, because less developed countries that are richer and/or higher-emitting will be held to account for climate change action in line with their level of wealth and/or emissions levels. From the above, it seems that a mixed approach would be perceived as fairest for all countries because it recognises the concerns of both less developed countries and developed countries better than either the PPP or ATP models. Considering the strengths of the mixed account, the chapter will now allocate responsibilities for adaptation and mitigation based on its own mixed model.

Allocating Climate Change Responsibilities to States – the PATP Model

The two existing mixed approaches assessed above, BPP and ATP*, both have important elements that should not be ignored when defining this account of responsibility. The most important element of the BPP approach lies in the suggestion for measuring both emissions and wealth levels in order to ascertain how much a country should contribute to climate change efforts. This ensures that responsibilities are allocated as fairly as possible, because it can take into account multiple categories of countries: high-emitting and wealthy, low-emitting and wealthy, high-emitting and poor, low-emitting and poor. However, it is important to qualify this index with the ATP* caveat that countries should not be pushed under a decent threshold of living by their climate change reduction efforts, which was illustrated to be a useful argument in terms of incorporating concerns of less developed countries, and in particular Least Developed Countries. In addition, the book has previously defended the importance of a decent standard of living and the ATP* ensures that this is protected. These two elements – the basic idea of using both wealth and emissions levels and the caveat about thresholds – will therefore be used as the building blocks of the mixed approach defended in this book.

This mixed approach will be referred to, for the rest of the book, as the Polluter's Ability to Pay approach, or PATP. Under the PATP model, countries that have high levels of emissions and high levels of wealth are responsible for reducing their pollution and paying for climate change adaptation and financing, and countries of low wealth and high pollution are responsible for reducing their emissions as best possible, as long as it does not push them under the threshold of a decent standard of living. Countries should have to pay for climate change costs and reduce emissions only when they are well off enough to do so. Furthermore, countries that have low emissions and low wealth should be excluded from responsibility, and those with low emissions and high wealth should be asked to contribute financially to adaptation and technology transfer costs but not lower their emissions. The PATP in this way captures the complexity of the category 'less developed country'. Although this category is very broad, the PATP model points out two important factors that can help to clarify which less developed countries may be more responsible for sharing the burdens of climate change than others: level of wealth and level of emissions. The PATP model can thereby accommodate the idea that some richer less developed countries must be held to account for their emissions, while at the same time stressing that Least Developed Countries should bear no responsibility until their level of wealth and emissions meets a level which implies responsibility.

The PATP, then, is a model of responsibility for climate change action that attempts to incorporate notions of fairness as explicated by less developed and developed countries in order to represent a fair allocation of climate responsibilities. The second demand of justice is therefore as follows:

2. The distribution of benefits and burdens in global climate change action must be based on the PATP model.

Of course, the PATP model cannot guarantee that the concerns of all countries are taken into account in climate change negotiations. It is merely a model of distribution that attempts to capture these concerns. Unfortunately, the reality is that less developed countries, and in particular Least Developed Countries, are often not included in climate change decision-making processes (Bulkeley and Newell 2010: 31). This is problematic not only in terms of fairly allocating responsibility, but also because less developed countries have certain key concerns which must form a substantial part of the climate change response, most notably the need for adaptation. If less developed countries are not part of decision-making procedures, these concerns may fall by the wayside. This will allow power disparities to permeate the climate policymaking process, virtually guaranteeing policy outputs that benefit the powerful and cost the powerless (Vanderheiden 2008: 253). For this reason, the evaluation of current practice in Chapters 5 and 6 will focus not only on how responsibility has been distributed between state actors, but also discuss how well less developed countries have been included when decisions about responsibility are made. For now, the current chapter turns to discussing whether non-state actors have the responsibility to lower emissions and/or pay for adaptation costs alongside state actors, whose responsibilities to do so have now been defined.

THE RESPONSIBILITIES OF NON-STATE ACTORS

When considering responsibility for climate change, states often come to mind first, because they actively work together to address the climate change problem through the United Nations Framework Convention on Climate Change (UNFCCC). The responsibilities of states to lower emissions and/or pay for adaptation measures has now been established, but the question of who else might be responsible for climate change action remains open. The Introduction of this book explained that climate change is an issue of global justice because it places 'everyone, everywhere' in a situation of mutual dependency: there is only one atmosphere, and multiple actors, within and outside of states, contribute to changes in global climate, albeit with varying effects in different places, regardless of where they are located (Harris 2010: 153). This seems to imply that climate change is as a cross-level distributive justice issue among all actors causing and suffering from climate change impacts (Harris 2010: 116).

It is therefore important to discuss who the actors causing climate change are and consider how they relate to those who are suffering from climate change effects, or to those who will suffer from climate change effects in future. Furthermore, it is important to consider whether these relationships imply responsibilities for those who cause climate change. In other words, what is needed is relational analysis.

Using the relational side of the climate justice account, the chapter will conduct such analysis here, and put forward that non-state actors should indeed be held responsible for lowering emissions and/or contributing to adaptation. This is because non-state actors, in the case of climate change, contribute to and perpetuate a wider system of injustice, which threatens the human right to health of present and future peoples. Non-state actors capable of lowering emissions and/or contributing to adaptation costs should therefore be responsible for doing so. In order to make this argument, the chapter first turns to discussing non-state climate relationships before turning to capable actors.

The Relationship between Those Who Cause Climate Change and Those Who Suffer from Its Effects

This final part of the chapter puts forward that non-state actors are harming climate change victims by perpetuating a wider system injustice that threatens their right to health, and should therefore be held accountable. However, there are those who argue the opposite: individual actors cannot cause moral harm, and should therefore not be held to account for their actions in the case of climate change. Walter Sinnott-Armstrong (2010: 334), for example, claims that although it is indisputable that some actors can affect emissions levels, it is not possible to assign these actors moral responsibility to refrain from emitting under any existing moral principles. He argues that although one act, such as driving for pleasure on a Sunday, may contribute to climate change, the act itself does not directly harm anyone, because climate change only happens when emissions accumulate over time (Sinnott-Armstrong 2010: 335). In this way, it is not clear which action does what harm and there is no direct link between action and harm, which means that the harm, indirect harm and harm contribution principle all do not apply. As Sinnott-Armstrong (2010: 277) explains, the small act of driving does not make climate change worse, as no individual person or animal will be worse off because of this one action.

However, even if it were accepted that there is no moral responsibility for individual actors in the case of climate change because individual acts do not cause harm, it is undeniable that non-state actors are causing harm to the climate collectively. In fact, there have been studies which reveal that non-state actors have significant impact on emissions collectively. Take individuals as an example. An average adult in a wealthy country emits around 800 metric tonnes of GHGs in their lifetime (Broome 2012: 74–8). This is equivalent to annual GHG emissions from countries such as Haiti, Albania and Armenia (CAIT 2017). To put it into global context, 800 metric tonnes are currently equivalent to about 0.02% of global emissions shares (CAIT 2017). Importantly, the affluent living in less developed countries are not far behind those living in developed countries in terms of average lifetime consumption. Individuals living in affluent less developed countries such as China and India account for over one fifth of the 'global consumer class', a number that is approaching 400 million and that exceeds the number of people living in Western Europe (Harris 2010: 124). The collective contribution of all affluent individuals is therefore staggering. A recent study conducted by Oxfam (2017) found that the top 10% of wealthiest individuals account for 49% of global 'lifestyle consumption' emissions.

Individuals are of course not the only non-state actors collectively making a substantial impact on global emissions levels. Consider corporations as an example. A recent study found that Shell emits more than Saudi Arabia, Amoco more than Canada, Mobil more than Australia, and BP, Exxon and Texaco more than France, Spain and the Netherlands (Bulkeley and Newell 2010: 2). The impact of the richest corporations, in particular, is astounding – the top one hundred are responsible for 52% of global 'industrial' GHGs since the industrial revolution (Climate Accountability Institute 2017: 5). It is therefore undeniable that these rich corporations make a significant impact on global emissions, especially collectively. Finally, consider cities. According to the C40 (2017), cities are home to half of the world's population, and consume over two thirds of the world's energy, cumulatively accounting for more than 70% of GHG emissions. According to Hoornweg et al. (2011: 207), if all production and consumption-based emissions that result from lifestyle and purchasing habits are included, urban residents and their associated affluence likely account for more than 80% of the world's GHG emissions. Again, it is the most affluent cities who are collectively contributing most. The world's fifty largest cities, with more than 500 million people, generate about 2.6 billion metric tonnes of GHGs annually, which is more than all countries combined, with the exception of the USA and China (World Bank 2017: 16). Cities, then, make quite a substantial impact on emissions, especially as a collective.

The above implies that it may be worthwhile to consider collective harm done by non-state actors, rather than focusing on the harm of individual actions. Interestingly, Sinnott-Armstrong (2010: 340) argues that collective principles do not apply to the case of climate change, because it is not immoral to do what others do as long as others' behaviour cannot be changed, since the harm will occur with or without a particular individual action. He explains that because an individual cannot change what the group does, it may be morally good or ideal to protest what the group does, but it does not seem morally obligatory (Sinnott-Armstrong 2010: 340). Sandberg (2011: 141) argues that Sinnott-Armstrong is mistaken, and that in the case of climate change 'we have a collective obligation to change our ways'. As Sandberg (2011: 242) puts it, 'all drivers and flyers ought to reduce this behaviour, because their collective [actions are] currently causing a threat of climate change'. In other words, although individual acts do not cause harm, collectively these acts do threaten individual actors, and this raises an obligation to change behaviour.

The moral basis for this obligation stems from the wider system of injustice that is being created through cumulative emissions, which threaten the human right to health. This system establishes a relationship between those who are suffering climate change effects and those causing these effects, because individual actors can affect the system by changing their emissions-producing behaviour. The ability to change the system implies moral responsibility, because it implies that individual actors play a collective role in imposing unjust systems on their most disadvantaged (and involuntary) participants by perpetuating and contributing to the system (Pogge 1989: 276). In the case of climate change, such an argument implies that although it may not be possible to argue that an individual non-state actor causes direct harm through an act such as driving, the actor is nonetheless part of, and indeed contributing to, a wider system of injustice that threatens the human right to health of the least advantaged. Because individual actors could affect the system through their actions, they have a moral responsibility to not perpetuate and contribute to the system. In this sense, Sinnott-Armstrong's dismissal of collective principles is arguably too hasty, as an existing collective moral principle, outlined above, clearly points towards a moral obligation of non-state actors.

Importantly, not all actors have the same moral obligations. The system of injustice is usually mostly imposed by advantaged, powerful participants (Pogge 1989: 276). It is these powerful actors who share a collective responsibility for the unjust system, because they most strongly uphold and perpetuate the system (Pogge 1989: 247). In the case of climate change, there are powerful actors who are upholding and perpetuating the system and putting the right to health at risk. These actors include wealthy individuals, corporations and sub-state actors such as cities, as discussed above. According to the model of collective responsibility outlined above, these actors must be held to account. Importantly, holding such actors to account would not merely contribute to lowering global emissions and/or raising money for adaptation, but would also subvert the unjust system, allowing space for this system to be questioned and criticised. This is crucial in the case of climate change, because a change in behaviour is urgently necessary from all capable actors in order to protect those suffering from climate change effects. The chapter therefore puts forward that capable non-state actors, including individuals, corporations, sub-state entities and international institutions, should be held morally responsible for lowering emissions and/or contributing financially to mitigation and adaptation efforts.

Allocating Climate Change Responsibilities to Non-State Actors

If capable actors are to be allocated climate responsibilities, it is worth outlining who these capable actors are, and what they are capable of. Let us begin by considering individuals. Of course, individual consumers are only able to choose between options which are available to them. Some options are simply not available, for example clean, efficient, affordable and safe public transport (Seyfang 2005: 297). In addition, individuals are often left out of decisions that make a major impact on emissions because these are made on a societal, not individual level (Seyfang 2005: 296). Decisions on building and maintaining roads, hospitals and schools and equipping the military, for example, account for half of all consumption in Western Europe (Seyfang 2005: 297). All of this implies that individuals do not have unlimited choices to lower their emissions and are locked into certain choices and consumption patterns. However, it would be difficult to argue that individuals do not have any impact at all. In fact, as was explained above, there have been studies which reveal exactly the opposite: many individuals do have a significant impact on emissions.

Furthermore, even if consumers are locked into certain choices, it would be impossible to argue that individual consumers have no way of altering their behaviour to affect emissions. Take one aspect of consumption as an example: food. About 25% of global GHG emissions come from food production (Vermeulen et al. 2012). Scarborough et al. (2014) conducted a study of British diets and found that the average daily GHG emissions in kilograms were 7.19 for high meat-eaters (over 100 grams a day), 4.67 for low meat-eaters (under fifty grams a day), 3.91 for fish-eaters, 3.81 for vegetarians and 2.89 for vegans. In other words, GHG emissions were twice as high for meat eaters than vegans. This means that those who already eat high levels of meat are capable of reducing their emissions significantly by reducing their meat and dairy consumption. Changing diet is most realistic when one has the means to do so – the capability to spend money on a nutritious vegan diet, for example. This differentiation in capabilities should not be ignored and must be considered when deciding which individuals should be held responsible for climate change action.

This type of capability as responsibility approach is often used in philosophy and has famously been defended by Peter Singer (1972: 231), who argues that if it is 'within our power to prevent something bad from happening, without thereby sacrificing anything of comparable moral importance, then we ought, morally, to do it'. Singer (1972: 231) illustrates his argument by explaining that if someone is walking past a shallow pond and sees a child drowning in it, they ought to wade in and pull the child out, even if this will mean getting their clothes muddy. The muddy clothes are not of comparable moral importance to the death of a child in this example. In the case of climate change, it is surely within the power of a rich individual to reduce their meat consumption without sacrificing something as morally important as the violation of the human right to health of millions. Being healthy enough to lead an adequate life is not morally comparable to the happiness that may result from eating meat for every meal. Health is fundamental to living an adequate life, whereas eating meat may bring short-lived satisfaction but is not essential to living a 'fully human' life. There are of course ways individuals can lower emissions – not flying as much, consuming local products, owning fewer cars, and so on. Wealthy individuals are more than capable of changing this behaviour (Harris 2010: 124). According to Singer's logic, wealthy individuals, who are capable of mitigating climate change and therefore protecting the right to health of potential climate change victims, should be held responsible for doing so.

If the moral responsibility of individuals can be established by considering their capability, then surely this logic can, and should, be extended to other actors. As was explained above, corporations and sub-state actors make a significant difference to emissions levels, especially collectively. Intuitively speaking, it seems that the argument from capability made by Singer and Harris could be applied to assign moral responsibility to these powerful actors. Cities, for example, can significantly reduce their emissions, even through simple means like switching to LED lights. The LED Lighting Project, in place in major cities such as London, New York, Hong Kong and Mumbai, has shown that switching to LED lights in cities can present energy savings as high as 80% (Climate Group 2012: 24). This is quite promising, especially if this kind of energy saving can be applied to all cities across the world. Street lighting accounts for 6% of global emissions levels, which is the equivalent of GHG emissions from 70% of the world's passenger vehicles (Climate Group 2014). If the LED Lighting Project is applied globally it would contribute to lowering the 6% of global emissions associated with street light down to 1.2%, because LED lamps are 80% more efficient than normal street lamps. In addition, cities who take ambitious action can act as a positive example and role models, and pressure their federal governments to take action on lowering emissions (Pattberg 2010b: 152). Beyond lowering emissions, cities can also disseminate knowledge and exchange best practices on issues ranging from energy-efficient buildings to water and waste treatment (Hoffman 2011: 12). All of this implies that cities are more than capable of lowering global emissions and contributing to the climate change response, without sacrificing something as morally important as the human right to health.

Similarly, corporations are capable of lowering emissions substantially – and in many cases, they are already trying to. IBM, the world's biggest computer services provider, cut its electricity use by 20% between 2008 and 2012. Similarly, FedEx, the world's largest overnight delivery company, is aiming for a 30% cut in aircraft emissions by 2020 from 2005 levels. These corporations are not alone – companies from Google to Walmart to Bank of America, among others, have pledged to power their operations with 100% renewable energy, with varying deadlines. In addition, beyond simply lowering emissions, groups of corporations can effectively institutionalise new norms at the transnational level; for example, the norm to disclose corporate carbon emissions (Bulkeley and Newell 2010: 87). These emerging norms are expected to motivate and facilitate meaningful dialogue among business actors, investors and the wider public to induce corporate responses to climate change (Bulkeley and Newell 2010: 16). Furthermore, it has been estimated that 98% of global investment and financial flows required to tackle climate change will need to come from the private sector, because the private sector develops and disseminates most of the world's technology (Bulkeley and Newell 2010: 88). For this reason, the private sector will play a significant part in implementing and financing individual governments' climate change policies. Finally, corporations can also directly donate to climate change efforts. IKEA, for example, has pledged to spend €1bn ($1.13bn USD) on renewable energy and steps to help less developed countries cope with climate change. All of this suggests that corporations are more than capable of contributing to climate change efforts without sacrificing something as morally important as the human right to health.

Simon Caney has followed up on the intuition that capable non-state actors should be held to account. He argues that all agents who are capable of reducing their emissions should be held accountable for doing so (Caney 2010a: 219). Importantly, Caney (2005: 755) suggests that if these actors are not only high emitters, but affluent, they have the additional capability of financially contributing to climate change efforts, such as scientific research into climate science, or adaptation measures, and should therefore be responsible for doing so. This argument for financial contribution is in line with capability logic – spending money on preventing future human rights abuses (specifically violations of the right to health in the case of climate change) is morally required of agents capable of doing so, as long as spending funds does not result in a moral harm equal to the violation of the human right to health. What Caney puts forward, then, is a model of responsibility where those with the capability of lowering emissions and/or paying for climate associated costs must be held responsible. In this sense, Caney is simply extending the capability argument in order to hold to account all capable actors.

Extending the argument from capability is especially useful in the case of climate change because it allows powerful actors, for example corporations or cities, to be held morally accountable for perpetuating the wider system of climate injustice. The most capable actors should be held to account for lowering emissions, but also for adaptation efforts, considering the importance of adaptation for the protection of the human right to health. Their responsibility must be allocated in line with their respective capabilities, as long as this does not imply sacrificing something as morally important as the human right to health. Thus, the third and final demand of climate justice is:

> 3. Capable actors, including individuals, firms, sub-state entities and international institutions, irrespective of the country they live or exist in, must be held responsible for lowering emissions and/ or contributing to adaptation efforts, in line with their respective capabilities.

Chapters 5 and 6 will evaluate to what extent multilateral and trans-national climate change responses have been able to hold these actors to account. In addition, these chapters will discuss to what extent Demands One and Two have been met by the global response to the climate change problem. First, this current chapter will sum up what has been argued in Part I of the book, now that it is drawing to a close.

CONCLUSION

The first part of this book focused on four questions raised by the climate change problem: who will be most affected, what exactly is at stake, what action must be taken in the face of climate change, and who should be responsible for this action. In answering these questions, the book has now set out a climate justice account that specifies what exactly is normatively expected of the climate change response, in the form of three distinct demands of climate justice. Demand One centres around a minimum set of actions which must be taken to protect the human right to health: keeping temperature changes at or below 2°C and prioritising adaptation alongside mitigation. The second and third demands centre on who is responsible for ensuring temperatures are kept to this minimum and/or who is responsible for paying for adaptation. Demand Two stipulates that the responsibilities of states should be allocated according to the PATP model. Finally, Demand Three specifies that capable actors, including individuals, firms, sub-state entities and international institutions, irrespective of the country they live or exist in, must be held responsible for lowering emissions and/or contributing to adaptation efforts, in line with their respective capabilities. These three demands are considered normative principles that must underwrite a more just global response to climate change.

The first part of the book has now laid the foundation for the assess-ment of current practice, by defining the scope and grounds of justice and explaining what climate justice demands. It was important to define these three elements of the climate justice account, because this has clarified what exactly is normatively expected from the global response to climate change. Now that this is clear, the assessment of this response can commence. The assessment will focus on both multilateral and transnational climate responses and aim to pinpoint what has gone right, what has gone wrong, and what has changed since the Paris Agreement was ratified. Part II will focus on both positive trajectories and current hindrances and comment on what the future of climate governance might look like, especially in terms of a just response to the climate change problem. The overall aim is to provide a comprehensive cosmopolitan assessment of climate change governance, both multilateral and transnational.

Part II

Assessing Climate
Governance

Chapter 4

BRIDGING THEORY AND PRACTICE

INTRODUCTION

Having developed a climate justice account in Part I, the book now turns to assessing the global response to climate change in Part II. This assessment will focus on both multilateral (state) and transnational (non-state) responses, and in doing so make sense of the 'big picture' of climate change (mis)management and the injustices that come along with it. Although some climate governance scholars are beginning to turn their attention to climate justice (Bulkeley, Edwards and Fuller 2014; Hale 2016; McKendry 2016), and some climate justice scholars have begun to explore and assess the role of non-state actors (Heyward and Roser 2016; Moss 2015; Maltais and McKinnon 2015), there is a lack of systematic research that normatively evaluates transnational climate change responses, or explores how these compare to multilateral efforts. In other words, there remains a significant gap between theory and practice. The book aims to bridge this gap, by illustrating that climate justice theory can be used to assess climate change governance, both multilateral and transnational. In order to begin bridging the gap, the current chapter explains exactly how this assessment will be conducted. The chapter is organised as follows.

First, the chapter provides an overview of global climate change governance, clarifying what aspects of the global response the book focuses on. Next, the chapter explains why actors who operate within climate change governance processes have a responsibility to act on the demands of justice defined in Part I of the book. The chapter grounds the responsibility of these actors in their capability to enable the three demands of climate justice by restructuring the social and political context to make this possible. Following this, the chapter will consider which actors are to be held most responsible if both multilateral and transnational climate change governance actors have the capability to restructure the context. Finally, the chapter develops a methodological framework for assessment. This framework is based on a four-point hierarchy that can be used to investigate to what extent global governance actors enable each demand of justice. In this sense, the current chapter sets out a plan for cosmopolitan assessment of climate governance, and thereby begins to bridge the gap between those who focus on climate justice and those who research the global response to the climate change problem. The assessment of climate change governance in Chapters 5 and 6 will further expand this bridge. Overall, Part II of this book aims to demonstrate that the bridging of the fields of climate justice and climate governance allows for a comprehensive normative understanding of the global climate change response that can underwrite future research and ultimately help to bring about a more just global response to climate change.

THE GLOBAL RESPONSE TO CLIMATE CHANGE

The global response to climate change is almost immeasurable, with institutions in place at the global, regional, national and local level. In order to narrow down this vast response, the book will focus on the global governance of climate change. Global governance has been prioritised because climate change is a transnational collective action problem that is unlikely to be resolved by action at the level of the nation-state (Bevir 2009: 8). For this reason, climate change action is coordinated largely at the global level, and the decisions made at this level have an effect on all those below it. In this sense, investigating regional, national or local governance may be difficult without an understanding of global institutional practice, because decisions at the regional or national level will have been influenced by global-level decisions. It is therefore worthwhile to assess global climate change governance as a priority. Furthermore, although assessing regional, national or local institutional practice is no doubt important and interesting, the book cannot hope to assess all aspects of the global response to climate change – it is simply too vast. It is therefore necessary to focus on one area of governance in order to be able to conduct a focused and coherent assessment.

The global governance of climate change can be defined as 'all purposeful mechanisms and measures aimed at steering the social systems towards preventing, mitigating, or adapting to the risks posed by climate change' (Stripple and Pattberg 2010b: 142). This broad definition captures the fact that global climate change governance is a complex array of many different types of institutions, organisations, regimes, regulatory bodies, decision-making procedures and actors. It is therefore challenging to pin down exactly what processes are occurring, who is responsible for what action and who has authority to act. This chapter attempts to provide clarity on all of three of these questions, beginning with the question of what kinds of processes are occurring. After this has been clarified, the chapter turns to explaining who is responsible for climate justice and discussing how authority might affect these responsibilities.

The Processes of Climate Change Governance

Although climate change governance is a vast arena of actors and institutions, these processes can be broken down into two broad categories common to global governance as a whole: multilateral and transnational climate change governance. Multilateral climate change governance refers to the state-led processes of climate change governance coordinated under the United Nations Framework Convention on Climate Change (UNFCCC). The UNFCCC is the only global multilateral institution tasked, by treaty, with coordinating global climate change action. It will therefore be the only institution assessed under the heading of multilateral climate change governance processes. The UNFCCC provides a framework to assess progress and to negotiate policy and international treaties. The main actors operating under the UNFCCC are states, referred to as Parties to the Convention, who meet annually at the Conference of the Parties (COP) to review progress on commitments and to update them in the light of the latest scientific advice. The ultimate objective of the UNFCCC is to achieve stabilisation of greenhouse gas concentrations in the atmosphere at a level that would prevent dangerous anthropogenic interference with the climate system (UNFCCC 1992: 9). At the annual COP, states assess current measures taken, and work towards developing treaties and implementation measures. The most significant outcomes of the COPs to date are the 1997 Kyoto Protocol, which committed a number of developed countries to GHG reduction and limitation targets, and the 2015 Paris Agreement, which focused on a global target of keeping temperatures below 1.5°C and ensuring countries set their own emissions targets in line with this aim.

Alongside the UNFCCC, there exists an array of non-state actors that are attempting to respond to climate change, including cities, multinational corporations, international organisations and individuals who coordinate their action globally. These actors promote policies, set standards and call for action both with and without the cooperation of states (Bevir 2009: 87). Climate change governance scholars often group these types of actors under a common umbrella term such as experimental governance (Hoffman 2011), networked governance (Stevenson and Dryzek 2014) or transnational governance (Bulkeley et al. 2015). The book will use only one term, namely 'transnational governance', in order to avoid confusion. The term should be taken to imply 'experimental' governance or 'networked' governance by those who refer to it as such.

Transnational climate change governance actors are not focused on a single outcome, such as a global treaty. Instead, transnational climate change governance initiatives push the global response to climate change in a number of directions, including energy efficiency, carbon markets, local adaptation and transformation of the built environment or transportation systems, among others (Hoffman 2011: 5). Most commonly, transnational climate change initiatives are involved in information sharing or voluntary goal-setting. In this sense, transnational climate change governance is a collection of individual initiatives which are decentralised and self-organised (Bulkeley et al. 2015: 156). Recent years have produced a 'Cambrian explosion' of networked climate change governance initiatives (Abbott 2012: 571). It is therefore challenging to discern the 'total universe' of networked climate change governance activities (Bulkeley et al. 2015: 19). In fact, the exact number of transnational climate change governance initiatives is unknown, although it is estimated that the numbers of projects in existence is in the thousands (Stevenson and Dryzek 2014: 87). Nevertheless, these processes can be categorised in order to provide an oversight of what is occurring.

One of the most common ways to categorise transnational governance processes is to group the activities into public, private and hybrid (Stevenson and Dryzek 2014; Stripple and Pattberg 2010; Bulkeley and Newell 2010; Bäckstrand 2008). Public transnational climate change governance processes include transnational municipal networks, networks of regional governments or bilateral agreements between subnational governments (Bulkeley and Newell 2010: 59). These public groups usually focus on meeting common goals. Examples include the C40, which is a network of the world's largest cities that aims to share best practices and develop collaborative initiatives on city-specific issues (Hoffman 2011: 19). Private transnational climate change governance processes, on the other hand, involve a variety of private actors, including corporate and civil-society sectors, who work together to define issues, set up rules to follow and ensure compliance to these rules is monitored (Bulkeley and Newell 2010: 65). Examples include the Verified Carbon Standard, a private institution that facilitates the exchange of carbon credits by 'eliminating the need for the purchaser to evaluate the merits of many different projects', and therefore playing a crucial role in enabling carbon markets to run (Stevenson and Dryzek 2012: 111).

Finally, hybrid transnational climate change governance processes include both public and private actors in various forms of collaborations (Bulkeley and Newell 2010: 62). This type of transnational climate change governance can involve a range of actors, including host governments, private investors, carbon brokers and non-governmental organisations. An example of a hybrid project is the Climate Group, which involves both public actors (for example Germany, California and London) and private actors (for example M&S, BP, HSBC, Shell), and aims to spread best practices and solutions (Hoffman 2011: 82). Grouping the processes of transnational climate governance into public, private and hybrid governance is common in climate change governance literature, and so Chapter 6 will employ these same categories to make sense of how transnational actors are faring in terms of enabling a just response to climate change.

The Importance of Assessing Both Processes

It is important for the book to assess both multilateral and transnational governance, because both processes play an important part in acting against climate change (Bulkeley and Newell 2010: 10). The multilateral regime has been hard at work since 1992, when the first COP was held. States have come together every year since then, producing both the Kyoto Protocol and the Paris Agreement. The Kyoto Protocol remains in force until 2020, at which time it is expected to be superseded by the recently ratified Paris Agreement. The multilateral regime is clearly not losing momentum. If anything, it is growing in importance, as more and more states commit to climate change targets. Assessing these processes therefore remains paramount for climate justice scholars, who have begun investigating these processes in recent years (Vanderheiden 2008; Harris 2010; Lawrence 2014). Questions to ask include whether the Paris Agreement presents a meaningful and significant step forward, whether allowing individual states to set their own emissions targets will increase participation in climate efforts and whether the newly conceived five-year review process will ensure that these voluntary targets are met. These questions are especially important while the particulars of implementing the Paris Agreement are still being discussed, because there is still time for change.

Although multilateral climate change governance processes are perhaps more public or familiar, 'we no longer observe a singular global response to climate change, and are instead observing multiple global responses', including transnational climate change governance responses (Hoffman 2011: 17). Furthermore, it has become clear that transnational actors have the potential to contribute to the climate change response in a significant manner (Hsu et al. 2015). Their role is especially important at this present moment because the Paris Agreement presents a significant shift away from a purely state-led approach. Under the post-Paris Agreement climate regime, transnational climate change actors are no longer seen as merely a 'helpful addition', but as a core element of the global response to climate change (Hale 2016: 14). In this sense, failing to explore and assess transnational governance processes alongside multilateral ones would ignore the complexities of the global response to climate change (Bulkeley and Newell 2010: 10).

Furthermore, while it may be difficult to research transnational climate change governance processes due to their diversity and complexity, their increasingly important role in the global response to climate change raises normative questions that must be answered. As a cosmopolitan climate justice scholar, one should be particularly interested in how transnational climate change responses compare to multilateral responses in terms of a just distribution of burdens, allocation of responsibility, protection of human rights and procedural justice. These questions remain largely unexplored by climate justice scholars. Hoffman (2011: 5) has argued that transnational climate change governance processes may represent the best hope for effectively responding to climate change. By contrast, Hale (2016: 20) wonders whether the pragmatic, problem-solving approach of transnational actors risks detracting from questions of climate justice. The question is, then, whether these transnational actors help foster and contribute to a just response to the climate change problem, or whether they will instead undermine it. For this reason, the book will assess both multilateral and transnational climate change governance. To aid this assessment, the chapter now turns to explaining why actors under both processes can be held responsible for a just response to climate change.

The Responsibilities of Climate Change Governance Actors

Chapter 3 established the responsibilities of state and non-state actors to lower emissions and/or contribute to climate change adaptation in order to protect the human right to health. If these individuals, corporations, sub-state authorities, international institutions and states are all responsible, the question is how these actors can meet these responsibilities. The 'political reality' of climate change is that actors will not automatically comply with their moral responsibilities (Caney 2014: 134). Thinking otherwise would be quite naïve, considering the slow pace of progress on climate change in the past twenty years. In this sense, it would be misguided to assume that state and non-state actors will meet their responsibility to protect the human right to health without any intervention. However, the 'institutional reality' of climate change is that there are some actors who are capable of ensuring compliance with moral responsibilities or at least enabling action in the first place, for example by setting out policies, creating accountability structures and facilitating implementation of climate policies (Caney 2014: 135). This is an interesting notion to consider, especially in relation to climate governance actors. Intuitively, multilateral and transnational governance actors seem capable of enabling state and non-state actors to meet their climate responsibilities. It is worth exploring this idea further, to establish whether climate governance actors are indeed capable, and whether this might imply moral responsibilities on their part.

The Capability to Enable Climate Change Action

Simon Caney (2014: 141) argues that the capability to enable climate change action implies moral responsibility: those with the power to 'make a valuable difference', and the power to 'compel or induce or enable' others to act have the moral responsibility to do so. Although Caney here refers to 'power', not 'capability', his argument seems to be an argument from capability. This is implicit in the fact that Caney refers to those responsible as 'those who can', and that he argues that the 'capacity' of responsible agents is variable (2014: 143). Furthermore, Caney footnotes Peter Singer, explaining that his 'underlying line of reasoning has a structural similarity to that advanced by Singer'

(Caney 2014: 142). Singer advocates a responsibility as capability approach, as was explained in the previous chapter. Interestingly, Singer himself argues that it is difficult to envisage 'any solution to climate change that does not require effective global institutions' (Singer 2006: 421). He also explains that although it will not be easy for global institutions to regulate climate change, the challenge nonetheless 'needs to be met' (Singer 2006: 421). Singer is making two points here. The first is that institutions may have the capability to meet the challenge of regulating climate change efforts, and the second is that it is morally necessary that they meet this challenge.

According to Singer's (1972: 231) own responsibility from capability argument, which asserts that actors capable of preventing moral harm must act as long as they do not sacrifice something of equal moral importance, this seems to imply moral responsibility for capable institutions in the case of climate change. If global institutions are capable of 'meeting the challenge of regulating climate change efforts' and this challenge 'needs to be met' then this implies moral responsibility for global institutions, especially because responding to climate change is necessary to prevent moral harm – the violation of the human right to health. Further evidence for this claim lies in Singer's optimism for the future, which he bases in the fact that 'the developed countries have signed the Kyoto Protocol, and are now discussing further steps that need to be taken' (Singer 2006: 421). Singer is seemingly putting faith in global institutions and their potential capability to deal with the climate change problem, which implies, according to his own logic, that these institutions have a responsibility to act on their capabilities. While Singer does not make the case for this responsibility explicitly, it is in line with his original argument and implicit in his work on the climate change problem. This suggests that the allocation of moral responsibility to climate change governance actors could rest on an argument from capability, as it does in Caney's account.

The question to ask, then, is whether climate change governance actors are indeed capable of enabling state and non-state actors to meet their climate change-related responsibilities. If they are, climate governance actors could be held responsible for doing so. To consider this question properly, it is important to remember that climate change actors operate in social, political and economic contexts. Action on climate change is often influenced by these contexts, for example by the context of Republican leadership in the USA, which has affected their membership of the Paris Agreement, or the context of increasingly cost effective green energy, which has affected the Indian government's decision to scale back plans for new coal plants in favour of solar energy farms.

According to Caney (2014: 135), it is possible to structure these social, economic and political contexts 'in ways which induce agents to comply with climate responsibilities'. Caney asserts that there are agents capable of doing exactly this, and that they should be held responsible for doing so. Restructuring the context is a notion which is left quite broad in Caney's work, and needs further exploration and unpacking. Caney explains that restructuring can include enforcing compliance through policy measures, incentivising actors by offering rewards for their action, creating norms which encourage compliance by making non-compliance seem unacceptable, undermining resistance to compliance through accurate media representation of climate change, and using civil disobedience to encourage governments to act (Caney 2014: 136–8). Although this list is helpful, these are broad, and therefore necessarily vague, sets of actions. To unpack the idea of 'enabling' action on climate change further, the chapter will focus on whether climate change governance actors are capable of enabling the three demands of justice set out in Chapter 3 by restructuring the context state and non-state actors operate within. If they are indeed able to do so, this would imply a moral responsibility on their part.

The Capabilities of Multilateral and Transnational Climate Change Governance Actors

In order to explore the capabilities of multilateral and transnational climate change governance actors, the discussion below will consider to what extent both sets of actors are capable of enabling each demand of justice in turn. Readers should keep in mind that states are the ultimate decision-makers in the UNFCCC. For this reason, assigning responsibility to actors under the UNFCCC implies assigning responsibility to states who are signatories to the UNFCCC. Actors working within transnational governance processes, by contrast, can vary from individuals to corporations to sub-state entities, so when reference is made to 'actors', these varied types of actors should be taken as implicit to the term. Transnational actors do not operate under a global treaty or strive towards a unified goal, which makes their capability to enable the three demands slightly less straightforward to assess than that of multilateral actors. For this reason, the chapter will use examples of transnational climate change initiatives to explain the claims made.

In addition, before the chapter turns to assessing whether climate governance actors are capable of enabling the three demands of justice, it should be stressed it is the capability of the actors within climate change governance which implies responsibility, and this responsibility is not affected by whether or not these actors are likely to enable actors to comply with their responsibilities. As David Estlund (2014: 117) has recently argued, the fact that an agent's action is unlikely, even extremely unlikely, does not entail that it is beyond the agent's ability. Estlund (2014: 119) explains that impossible implies that an agent cannot perform an action. Improbable, on the other hand, implies that an agent is capable of an action, but there is no chance that she will do it, that the (objective) probability of her doing it is zero. He suggests that even a zero probability that an action will be performed does not entail inability (Estlund 2014: 119). What Estlund is implying is that probability of an action does not affect the capability of performing an action. Or, as Estlund (2014: 122) puts it, 'the likelihood that a person will not behave in a certain (entirely possible) way simply does not bear on whether they morally should: it is not a fact that has that kind of moral significance'. Therefore, even if climate change governance actors are unlikely to live up to their responsibility, say for example because of a lack of political will, this does not take away from their moral responsibilities. With this clarified, the chapter turns to exploring whether multilateral and transnational actors are capable of enabling all three demands of climate justice.

Demand One states that global temperature changes must be kept at or below 2°C and adaptation must be prioritised alongside mitigation. Actors under the UNFCCC have the capability to enable the first part of Demand One because these actors can restructure the context by setting emissions level targets. Keeping temperatures below 2°C requires substantial emissions reductions: 40–70% by 2050. The UNFCCC are explicitly tasked with creating global treaties and agreements in order to 'to achieve stabilisation of greenhouse gas concentrations in the atmosphere at a level that would prevent dangerous anthropogenic interference with the climate system' (UNFCCC 1992: 9). In other words, the UNFCCC has been set up exactly for the purpose of developing policies to mitigate emissions. States signatory to the UNFCCC therefore clearly have the capability of restructuring the context to ensure that the first part of demand is met, for example by setting out regulations, ensuring that states meet their emissions targets or defining global goals for temperature limits. This context is well under way to being created – the Kyoto Protocol set out specific emissions targets (at least

18% below 1990 levels in the commitment period 2013 to 2020) and the Paris Agreement sets out a 1.5°C temperature limit.

Furthermore, multilateral actors can also restructure the context to ensure that adaptation is prioritised alongside mitigation, thereby enabling the second part of Demand One. Article 4.4 of the Convention calls on developed country Parties to 'assist the developing country Parties that are particularly vulnerable to the adverse effects of climate change in meeting costs of adaptation to those adverse effects' (UNFCCC 1992: 14). From this responsibility to support adaptation stem two further responsibilities: financial and technological transfer. Article 4.3 of the Convention states that 'the developed country Parties ... shall provide new and additional financial resources to meet the agreed full costs incurred by developing country Parties' and Article 4.5 states that 'the developed country Parties ... shall take all practicable steps to promote, facilitate and finance, as appropriate, the transfer of, or access to, environmentally sound technologies and know-how to other Parties, particularly developing country Parties, to enable them to implement the provisions of the Convention' (UNFCCC 1992: 14). These Articles illustrate that the members of the UNFCCC aim to help less developed countries with adaptation, specifically through financial transfer, and transfer of technologies, which should assist these countries to develop cleanly. In this sense, the UNFCCC is set up specifically to develop and update adaptation-related policies at the COPs every year, and thereby restructure the context and enable action on adaptation. Again, this context is well under way to being created, with the Kyoto Protocol establishing specific mechanisms which serve to enable technological and financial transfer, including the Adaptation Fund, and the Paris Agreement putting in place an Adaptation Committee. In this sense, the UNFCCC is able to create a context which allows the second part of Demand One to be acted upon.

Transnational climate change governance actors, on the other hand, are capable of enabling the first part of Demand One because these processes include projects which have significant potential to substantially lower global emissions, creating a context in which the first part of Demand One can be enabled. Take for example the C40, a network of the world's largest cities aiming to share best practices and develop collaborative initiatives on city-specific issues in order to make implementing the global response to climate change more feasible (Hoffman 2011: 95). The C40 (C40 2017b) promises to have 'a meaningful global impact in reducing both greenhouse gas emissions and climate risks'. There are a number of reasons why the C40 is capable of enabling the lowering of global emissions. For one, only twenty of the ninety C40 cities are held to account for emissions reductions under the Kyoto Protocol. This is promising in terms of creating a context of lowering emissions. If the seventy cities not held to account under the Kyoto Protocol lower their emissions because of the C40, this will add to multilateral governance efforts. Second, the C40 encompasses a large part of the global population: representing over 600 million people, which amounts to one in twelve individuals worldwide (C40 2017b). The C40 therefore has the capacity to make a very real impact on climate change by creating a broad network, or context, within which emissions can be reduced. It seems the creation of such a context is well under way. Cities who are members of the C40 are three times more likely to take action on climate change, with over 10,000 climate change initiatives currently in place across the C40 (C40 2017a). This is of course just one example of a transnational climate change governance initiative. Further examples will be explored in Chapter 5. For now, the example of the C40 serves to very briefly illustrate the capability of transnational actors to enable the first part of Demand One.

Furthermore, transnational climate change actors are capable of enabling the second part of Demand One, because there are transnational climate change initiatives that focus specifically on adaptation. For example, Asian Cities Climate Change Resilience Network (ACCCRN) aims to strengthen the capacity of over fifty rapidly urbanising cities in Bangladesh, India, Indonesia, the Philippines, Thailand and Vietnam to survive, adapt and transform in the face of climate-related stress and shocks (ACCCRN 2017). The ACCCRN can enable the second part of Demand One in two ways. First, the ACCCRN creates a context where this is possible. Its members focus on helping individuals and organisations build climate change resilience for poor and vulnerable people by fostering partnerships and collaboration, creating a context where adaptation is prioritised. Second, the ACCCRN plans to expand this context. Its members aim to build a larger coalition to drive the capacity and action needed for climate change resilience in the region (ACCCRN 2017). In this sense, the ACCCRN provides an example of the role transnational climate change governance actors can play in creating and expanding a context where adaptation is prioritised alongside mitigation. Of course, this is just one example of transnational climate change governance and adaptation – further examples will be explored in Chapter 5. For now, the ACCCRN serves to demonstrate that transnational climate change actors have the capability of enabling the second part of Demand One. The chapter has now established the capability of both multilateral and transnational actors to enable Demand One, and so turns to the second demand.

Demand Two states that the distribution of benefits and burdens in global climate change action must be based on the PATP model, which stipulates that states are to be held to account in line with their emissions and wealth levels.[1] Both sets of climate governance actors are capable of enabling this demand. Actors under the UNFCCC have the capability to enable Demand Two because the UNFCCC has been charged with drawing up global treaties to, among other things, regulate the distribution of the benefits and burdens of climate change. For this reason, the actors operating under the UNFCCC have the capability to change the context by designing global treaties to be based on the PATP model, therefore enabling the realisation of this model. In fact, the idea of 'common but differentiated responsibility' (CBDR), defended in the original Convention, is compatible with the PATP model. Both PATP and the CBDR insist that duties fall on all, and yet both also insist that different demands can be made of different parties. Furthermore, CBDR establishes that the duties to which a party is subject depend both on what they have done and what they are able to do, which is what the PATP establishes because the model incorporates both emissions and wealth levels. In other words, the CBDR seems flexible enough to accommodate the notion of a PATP distribution. This is important, because this flexibility allows for room for the creation of a context where Demand Two can be met. This will be further discussed in Chapter 5.

Like multilateral actors, transitional actors are also capable of creating a context where Demand Two is enabled. In terms of ensuring that benefits and burdens are distributed according to the PATP model, transnational actors can structure the context through information sharing or campaigning on the subject, which could build support for a PATP. Transnational governance projects often involve campaigning and advocacy, and actors within these processes frequently attend the annual UNFCCC COPs in the hope of influencing decision-making. In addition, transnational governance projects can include states which are high-emitting and/or wealthy and not currently held to account under the Kyoto Protocol, therefore creating a context for these states to meet their responsibilities. For example, the Asia-Pacific Partnership on Clean Development and Climate (APP), which has now come to an end, included Australia, Canada, China, India, Japan, South Korea and the United States of America as members. With the exception of Australia, none of these states were held to account for lowering emissions and/or contributing financially under the Kyoto Protocol at the time of the APP. The APP created a context where wealthy and high-emitting states were, albeit voluntarily, held to account for their responsibilities, which is required under the PATP model. This is of course only one example, but it is illustrative of the fact that transnational climate governance actors can create a context which enables the PATP. Further examples will be discussed in Chapter 6. The current chapter has now established the capability of both multilateral and transnational actors to enable Demand Two, and so turns to the final demand.

Demand Three states that capable actors, including individuals, firms, sub-state entities and international institutions, irrespective of the country in which they live or exist, must be held responsible for lowering emissions and/or contributing to adaptation efforts, in line with their respective capabilities. Both multilateral and transnational actors are capable of enabling this demand by restructuring the context to make this possible. Actors under the UNFCCC have the capability to enable Demand Three of justice because the UNFCCC is tasked with creating regulations around climate change and can therefore create a context within which actors can meet their responsibilities. The UNFCCC creates regulations that assign a certain emissions target per state, and this in turn allows states to regulate the emissions of individuals, corporations and sub-state entities within their borders. The UNFCCC also has the capability to set up fundraising targets for which states could raise money through taxes or fines on individuals, sub-state entities and corporations. In this way, actors under the UNFCCC have the capability to enable Demand Three, because these actors can design global treaties and create a context that allows non-state actors to meet their responsibilities to lower emissions and make financial contributions. This is especially true now that the Paris Agreement has been ratified. Although a groundswell of non-state climate action has been growing over time, the Paris Agreement seeks to bring this phenomenon into the heart of the new climate regime (Hale 2016: 12). In this sense, actors under the UNFCCC are more capable of creating a context where non-state actors can be held to account than ever before.

Transnational climate governance actors are also capable of enabling Demand Three, because transnational governance initiatives create a context which allows firms, sub-state entities, international institutions and individuals to act on climate change. By taking part in transnational initiatives, non-state actors are, albeit mostly voluntarily, being held responsible for climate change action, both in terms of lowering emissions and making financial contributions to projects that fight against climate change. Take for example the Climate Group, which aims to raise funds and implement green technology across the world. More specifically, the Climate Group aims to create a clean revolution through the rapid scale-up of low carbon energy and technology (Climate Group 2017a). The Climate Group is made up of over one hundred major corporations, sub-national governments and international institutions. The Climate Group claims that the combined revenue of its members is estimated to be in excess of $1 trillion USD (Climate Group 2017b). This is promising in terms of possible financial contributions to the climate change cause, because it creates a context where wealthy actors can be held to account for contributing financially to the cause.

One project that has benefited from this financial contribution is the 'LED Lighting Project', which took place in major cities such as London, New York, Hong Kong and Mumbai. This project has demonstrated that switching to LED lights in cities can present energy savings as high as 80% (Climate Group 2012: 24). Projects such as these create a context where wealthy actors can contribute to climate change costs and help to lower global emissions by spreading green technology. Through initiatives like the Climate Group, actors involved in transnational climate change governance processes are creating a context that allows non-state actors to act on their climate responsibilities. For this reason, the actors involved in transnational governance processes are capable of enabling Demand Three. Chapter 6 will explore further examples to illustrate this capability and assess to what extent Demand Three is being enabled by transnational actors.

The chapter has now discussed how each demand of climate justice can be enabled by both actors working with both state and non-state climate change responses. The chapter's illustration of capabilities implies, under the capability model defended above, that both multilateral and transnational climate change governance actors have the moral responsibility to enable the three demands of justice by restructuring the context so that these demands can be met. Now that the responsibility of multilateral and transnational actors has been established, the chapter turns to exploring whether either set of actors is more responsible for enabling the three demands of justice explicated in this book. This is imperative to explore because responsibilities must be clearly defined and laid out before the extent to which they are being met can be assessed.

WHO IS MOST RESPONSIBLE FOR ENABLING CLIMATE JUSTICE?

The chapter has so far established that multilateral and transnational climate change actors are capable of enabling the three demands of climate justice defined in Chapter 3 and therefore have the moral responsibility to do so. Nevertheless, there is an important difference between these actors that affects their responsibilities: the difference in their source of authority. The concept of authority in global governance has been explored by a number of different scholars in recent decades, resulting in various definitions and conceptions of what authority is and how it is derived. Authority can, for example, be defined as getting an actor to do what they would not do otherwise, creating new preferences in actors who were previously indifferent or at odds, or even mobilising new constituencies for political action (Avant et al. 2010: 10). For others, authority simply implies power: power to take the lead, power to make decisions and power to implement policies (Whitman 2009: 14).

There are also various conceptions of how authority can be derived. For example, one of the most formal means of obtaining authority in global governance is through legal authority (Pattinson 2007: 302). Actors in the UNFCCC possess this type of formal authority because the UNFCCC is based on an international treaty (the Convention), which is considered the most robust form of international law (Lawrence 2014: 103). Furthermore, the treaties, or protocols, that the actors in the UNFCCC decide upon are usually assumed to be international law (Lawrence 2014: 103). This provides actors under the UNFCCC with international legal status. A further formal manner of obtaining authority is through delegated authority from other authoritative agents (Avant et al. 2010: 11). The UNFCCC is made up of authoritative states, and these states delegate their authority onto the UNFCCC. The state is unquestionably an authoritative agent in global governance, largely because its leaders hold a democratic mandate from the people (Bell and Hindmoor 2009: 13). In this sense, the UNFCCC derives delegated authority from its memberships. Finally, a third formal source of authority is the mandate to act (Pattinson 2007: 302). The Convention has been ratified by 195 states, encompassing and representing a large proportion of the global population.

This implies that the UNFCCC has a global mandate to act on climate change, providing it with authority.

Actors operating within processes of transnational climate change governance, on the other hand, do not derive authority in a formal manner. Transnational climate change governance processes do not operate under a legally ratified international treaty and therefore do not have the legal status to act on climate change. Actors in transnational climate change governance processes also do not derive authority from the delegation of state authority, as the actors in these processes are almost exclusively non-state actors. Furthermore, transnational governance processes cannot be said to have a global mandate to act, because the projects taking place within these processes are not operating under a single global treaty or are backed by a global decision-making procedure. For this reason, it can be said that actors involved in transnational governance processes have no formal authority to act on climate change (Stevenson and Dryzek 2014: 87).

However, transnational actors can derive authority in less formal ways. For example, informal authority can be derived from filling a recognised gap in regulation (Stevenson and Dryzek 2014: 87). This is especially true if a community agrees that a task must be accomplished (Avant et al. 2010: 13). The international community has established climate change as an urgent concern, by setting up a multilateral institution tasked with addressing it and charging an international body of scientists to research the problem and provide guidelines for action. This indicates that the international community has agreed that there is a task to be accomplished, namely to address climate change. Actors involved in transnational climate change governance processes act to satisfy the international community's preferences by acting as if climate change is a problem and attempting to provide solutions to the problem. Furthermore, they can fill gaps in regulation by including previously unregulated actors, holding responsible non-state actors to account, and contributing to emissions reductions and adaptation efforts, as was outlined above. In this sense, transnational climate change actors derive their informal authority by acting on climate change in ways multilateral actors either have not, cannot or will not. Informal authority, then, seems to be very much about capacity to act, rather than having an official mandate or being directed by authoritative agents such as states, as is the case for actors in the UNFCCC.[2]

The difference between formal and informal authority to act on climate change has implications for the moral responsibilities of multilateral and transnational climate change actors. To explain these implications, consider the following thought experiment. Imagine there exists a pool, which has a lifeguard specifically assigned to save individuals from drowning. By some terrible coincidence, three people suddenly begin to drown simultaneously in this pool. The lifeguard cannot possibly save all three people, because they are at opposite ends of the pool. The lifeguard has the formal authority to act, because she has been hired to save individuals from drowning. She has also signed a legally binding contract of employment in which she has agreed to save drowning persons. The lifeguard has been trained to save lives, and she possesses a buoy to help keep an individual afloat. She is arguably the most authoritative agent at the pool to intervene in a drowning and is therefore the most responsible agent in case of a drowning, because formal authority has been vested in her to do so.

However, the lifeguard cannot save all three individuals. Imagine now that the pool has bystanders, who are aware of the drowning people. Some of these bystanders know how to swim, and have been trained in basic first aid, including CPR. In other words, these bystanders have the capability to save drowning persons, but no formal authority to do so. According to the capability argument made by Singer, the capable bystanders have the moral responsibility to save the drowning individuals the lifeguard is unable to save. This is because they have the capability to prevent a moral harm without sacrificing something equally morally important: they can prevent a death of an individual without putting their own lives in danger due to their capability to swim and perform CPR. In other words, even in the absence of authority, these capable actors are morally responsible, and should act accordingly. There seems to be no moral difference between the responsibility of the lifeguard and the capable bystanders in a scenario where the lifeguard cannot meet her responsibilities.[3]

Further to this, now imagine a second scenario where only one person is drowning. In this scenario, the lifeguard had the capability to save the individual, and the authority to do so, but is unwilling to act, because she is feeling tired, or lazy, or maybe does not care for the person who is drowning. According to the capability argument, this unwillingness to act would not take away from her moral responsibility to act. As was explained above, it is not the probability, but the capability of an actor which implies moral responsibility. Probability is not morally relevant, and so the fact that the probability of the lifeguard leaving her chair is low because of her lack of will to act does not affect her moral responsibility. Importantly, if the lifeguard does not, due to her lack of will, act on her responsibility to save the drowning individual, then the capable bystanders would have a moral responsibility to intervene, because the person charged with the authority to act would be failing to save the drowning person, as in the case of the three drowning persons above. In this scenario, as in the one above, the bystanders must jump in to save the drowning person, because of their capability to do so. It is not morally relevant to the bystanders whether the lifeguard will not or cannot save a drowning person. In both cases, the bystanders must intervene. The only case where the bystanders would not have to intervene is if the lifeguard saves the individual from drowning. In this case, the bystanders have no moral responsibility to help, because they are not capable of saving the individual since the individual has already been saved.

Actors in the UNFCCC are comparable to the lifeguard in the pool scenario. Although climate change is not as immediately threatening as a person drowning in front of a lifeguard, the human right to health of millions, if not billions, is very clearly under threat. The sense of urgency for immediate action is based on harm that will occur in the future, due to our capacity for rational foresight (as explained in Chapter 3). Actors in the UNFCCC have the formal authority to act on climate change, as was explained above. These actors therefore have primary responsibility to act before actors who do not have the authority to act, much like the lifeguard has the primary responsibility to act before bystanders do. However, if the actors in the UNFCCC are unable, or unwilling, to fulfil their responsibility to enable the three demands of climate justice, just as the lifeguard is unable or unwilling to save the drowning people in the above scenarios, then the bystanders with capability but no formal authority are morally required to step in.

In the case of climate change, the role of the capable bystander is played by transnational climate change governance actors, who, although capable of enabling states, individuals, corporations, sub-state authorities and international institutions to act on their responsibilities, do not possess the formal authority to do so. Their capability gives them the informal authority to act, as was explained above, but even more importantly, the capability to enable the three demands of justice means they also have the moral responsibility to act, just as the bystanders who are trained in CPR have the responsibility to save the drowning individuals. So, if the actors in the UNFCCC are unable or unwilling to meet their responsibilities to enable the three demands of justice, then other actors capable of this, specifically actors involved in transnational climate change governance processes, are morally required to act.

It should be noted that the presence of other capable actors at the pool, or in the realm of climate change governance, does not diminish the responsibility of the lifeguard, or indeed of actors working within the UNFCCC. The presence of capable actors does not let the lifeguard or actors in the UNFCCC 'off the hook' in terms of moral responsibility. Actors under the UNFCCC are under a moral obligation to act, based on their capability to act. The presence of other capable actors does not diminish this moral responsibility. Actors in the UNFCCC must act on their responsibility until it is fulfilled, and they can no longer contribute to enabling a condition of justice in the case of climate change, for example if the UNFCCC were to disband.

Similarly, actors involved in transnational climate change governance processes have the responsibility to enable a condition of justice as long as they are capable of doing so. The only possibility of escaping moral responsibility for transnational climate change governance actors is if they become unable to enable the three demands of climate justice, or if actors under the UNFCCC were to fully enable all three demands of justice, and a condition of climate justice was therefore achieved. In this case, transnational climate change governance actors would no longer be under a moral obligation to help, because they would no longer be capable of enabling a condition of justice, as this would already have been achieved by actors in the UNFCCC. This is comparable to the argument that the bystanders at the pool would have no moral responsibility to save a drowning individual if the lifeguard saved the individual, since the individual would already have been saved. Actors cannot be morally obligated to do what they cannot do (Lawford-Smith 2011: 456). This is the logical inverse of the capability argument. Only in the absence of capability – in the case of this book, the capability to enable a condition of justice – is there absence of moral obligation. For this reason, although actors under the UNFCCC have the formal authority to act and are therefore most responsible for enabling the three demands of climate justice, as long as actors under the UNFCCC are failing to meet their responsibilities, actors involved in transnational governance processes must step in, much like the bystanders at the pool. Now that the moral responsibilities of multilateral and transnational climate change actors have been set out, the chapter turns to explaining how the remainder of the book will assess to what extent these responsibilities are being met.

ASSESSING THE CLIMATE CHANGE RESPONSE

The chapter has so far explained what aspect of the climate change response the book will focus on and made the case for the moral responsibility of multilateral and transnational climate change actors to enable a condition of justice in the case of climate change. The two chapters which follow, Chapters 5 and 6, will assess to what extent these actors are meeting their responsibilities. The book will now briefly set out a methodological framework to clarify how the climate change response will be evaluated. The chapter has defended the idea that multilateral and transnational climate change governance actors must enable a condition of climate justice by restructuring the context so that the demands of climate justice can be met. The question to ask, then, is how to assess whether the context is being restructured.

The book will focus on one particular aspect of 'restructuring the context' – examining the policies which have been promised and which have been put in place by climate change governance actors. Global governance actors respond to the climate change problem in an eye-watering amount of different ways. The policies that they have set out, or are planning to set out, provide a category that can capture this variety without being overly narrow. 'Policies' is here used in a broad sense: the term captures enforcement mechanisms, incentives, creating norms, undermining resistance to effective climate policies, as Caney (2104: 136) suggests, and other policies such as regulation for emissions or adaptation finance, policies on inclusiveness or policies on ensuring compliance. These types of policies have the potential to ensure that the institutional context is structured to keep emissions in check and prioritise adaptation, thereby protecting the human right to health. The category of 'policies', then, is very wide, and allows the assessment to focus on a number of different ways the context is being restructured to enable a condition of climate justice.

Of course, there are some issues with examining existing policies, because climate change policies can often contain ambiguities, which could be interpreted as creating a context that is conducive to enabling a demand of justice, but do not imply that a condition of justice is guaranteed. During negotiations, the Parties of the UNFCCC often seek flexible language to accommodate the diverging positions of parties (Stevenson and Dryzek 2014: 70). Common ambiguities include the frequent use of 'shall' instead of more peremptory words like 'will' or 'must' (Stevenson and Dryzek 2014: 70). This can be problematic, because these ambiguities imply that policies can be interpreted in different ways, which calls into question what the actors under the UNFCCC have committed to, and what these commitments will mean in practice. Although it may appear that the commitments or ambitions of the actors in the UNFCCC are attempting to protect the right to health, for example, this does not guarantee the protection of said right. On the transnational side, the problem is less about ambiguity and more about lack of evidence for bold aims and promises. Most transnational climate change governance initiatives are relatively new and are experimental in the sense that we cannot yet be sure how they will turn out (Hoffman 2011: x). In addition, transnational initiatives are not monitored under a consistent or universal framework, which means evidence of concrete measurable impacts is scarce (Bulkeley et al. 2015: 161). The assessment of individual examples in Chapter 6 therefore must often rely on self-reporting, which may be biased. These problems of ambiguity and measurement will be further discussed in Chapters 5 and 6, respectively.

For now, the current chapter will outline a four-point hierarchy which will be used to explore to what extent multilateral and transnational climate change governance actors enable a condition of justice by restructuring the context to allow for this.

The Four-Point Hierarchy

1. Actors in the institution enable the demand of justice – the demand of justice is unequivocally fulfilled in its entirety.
2. Actors in the institution are consistently working towards enabling the demand of justice – the demand of justice is not yet fulfilled, but there are policies in place which are consistently leading towards this goal.
3. Actors in the institution have promised to begin working on enabling the demand of justice in the future – no policy has been adopted, but there is the potential for the creation of policy in order to consistently work towards enabling the demand of justice.
4. Actors in the institution do not enable the demand of justice – there has been no promise or attempt to enable the demand of justice and there are no policies in place.

Chapters 5 and 6 will make use of the above four-point hierarchy to evaluate to what extent multilateral and transnational climate change governance actors enable the three demands of justice set out in this book. Enabling a demand of justice is at the top of the hierarchy, because the demands of justice must be fulfilled in order to meet a condition of justice in the case of climate change. The fulfilment of a demand of justice is therefore the goal to aim for, which is why it sits at the top of the hierarchy. The hierarchy is important because it allows for the assessment of the extent to which climate change actors are able to meet the moral parameters set out in this book. If current practice fails to meet these moral parameters, the book can point to this, and discuss the hindrances which are in place at the institutional level. The objective of using the above hierarchy, then, is to make normative judgements about the global response to climate change. In this way, the book can provide detailed analysis that can underwrite future research and ultimately help to bring about a more just global response to climate change.

Finally, it should be noted that even if the assessment of climate change governance reveals that the demands set out in this book are not enabled, the demands will not be revised. Instead, the book will point out the discrepancies between the global response to climate change and the normative demands made in this book. Therefore, assessing the climate change response does not imply compromising the demands of justice to fit with the reality of current practice. The demands of justice are the goal towards which the global response to climate change should strive, and the assessment in Chapters 5 and 6 will reflect this. As Estlund (2014: 115) argues, 'the truth about justice is not constrained by considerations of the likelihood of success in realising it'. The reality of climate change governance does not affect the climate justice position defended in this book. For this reason, even if none of the demands of justice are enabled, it is still important to explain how the global response to climate change could move towards enabling climate justice in future.

CONCLUSION

This chapter served as a conceptual introduction to the assessment of climate change governance that follows in Chapters 5 and 6. It was put forward that both multilateral and transnational climate change actors have a moral responsibility to enable a condition of climate justice, due to their capability of creating a social and political context within which the three demands of justice defended in this book can be met. Moreover, the chapter argued that the actors under the UNFCCC have formal authority to act and are therefore more responsible for enabling a condition of justice in the case of climate change. However, it was explained that this does not diminish the moral responsibility of other actors, specifically those involved in transnational governance processes, if the actors under the UNFCCC should fail to enable the three demands of justice. Finally, the chapter illustrated how the book will conduct its assessment of climate change governance by setting out a methodological framework based on a four-point hierarchy. The book now turns to the climate-justice-focused assessment of multilateral and transnational climate change governance in Chapters 5 and 6 respectively.

NOTES

1. For a discussion of the PATP, see Chapter 3.
2. It is worth noting that actors in the UNFCCC may derive authority from the informal sources described above, on top of the formal sources of authority outlined above. This would be unsurprising, as most actors in global governance are authorised by a mixture of sources (Avant et al. 2010: 18). However, this does not change the argument about actors in the UNFCCC possessing formal authority while transnational actors do not.
3. This type of argument has been made in relation to humanitarian intervention. Pattinson (2008: 265) argues that legal authority to act does not make a moral difference in a humanitarian emergency such as genocide. In this type of scenario, the most effective actor (most capable of tackling the massive violation of human rights) must act, even in the absence of formal authority. Pattinson's reasoning for this is that humanitarian interventions entail mass human rights violations, which must be brought to an end sooner rather than later (Pattinson 2008: 265).

Chapter 5

ASSESSING MULTILATERAL CLIMATE GOVERNANCE

INTRODUCTION

The second half of this book focuses on the cosmopolitan assessment of the global response to climate change, both multilateral and transnational. The previous chapter argued that global climate change governance actors must be held responsible for enabling a condition of justice in the case of climate change and set out a methodological framework to assess to what extent these actors meet their responsibilities. It was explained that both multilateral actors under the United Nations Framework Convention on Climate Change (UNFCCC) and non-state actors involved in transnational climate change governance processes have a moral responsibility to enable a condition of justice, due to their capability of restructuring the social and political context so that the three demands defended in this book can be met. However, it was also argued that multilateral actors have formal authority to act and are therefore most responsible for enabling a condition of justice in the case of climate change.

The purpose of the current chapter is therefore to assess to what extent multilateral actors enable the three demands of justice developed in Part I of the book. Further to this, the chapter aims to provide a broad-brush overview of multilateral climate change governance. This will allow readers to develop an understanding of the multilateral climate change response, which has been under way since 1995. Taking each demand of justice in turn, the chapter will focus on the normative commitments made in the Convention as well as assessing the policies set out in the Kyoto Protocol and examining what has been achieved so far by multilateral actors. Following on from this discussion, the chapter will consider to what extent the Paris Agreement presents a change from existing policies. In this way, the chapter provides a historical overview of multilateral climate change action, as well as looking to the future. The final part of the chapter summarises the findings made, discussing what to look out for in the run-up to the Paris Agreement being implemented. This will be expanded on in the conclusion of the book.

MULTILATERAL CLIMATE CHANGE GOVERNANCE: A JUSTICE-BASED EVALUATION

Before the assessment of multilateral climate change governance can commence, the chapter must briefly explain how this assessment will be conducted. The assessment will proceed in three parts, focusing on one demand of justice at a time. In each part, the chapter will consider the promises made in the Convention before assessing the Kyoto Protocol, current action on climate change and finally the Paris Agreement. The Convention is an international treaty that aims to guide multilateral action on climate change. All decisions made at the COPs and all Protocols developed, including the Kyoto Protocol and Paris Agreement, are based on this original treaty, which has been ratified by 195 countries. The Kyoto Protocol was signed by more than 150 countries in 1997, and presents the current 'plan of action' of multilateral climate change governance actors. The first commitment period of the Kyoto Protocol covered the years 2008–12, with the second commitment period (2013–20) still in force today. The Paris Agreement, ratified in 2016, is widely expected to come into force once this second commitment period has ended. In this sense, the Paris Agreement represents the future of multilateral climate change policy.

It is important to evaluate normative commitments, current policies and future plans of the UNFCCC in order to gain a fair assessment of multilateral governance. Assessing both normative commitments and policies (present and future) is important because it helps to capture the ambitions of actors under multilateral governance as well as how these actors meet these ambitions in practice. In fact, doing so reveals that actors in the UNFCCC often set out normative commitments that are very much in line with the demands of justice defined in this book. And yet, the current policies and achievements of multilateral actors do not suggest the demands of justice are fully enabled. It is important to investigate the reasons behind this phenomenon, and this would not be possible without an understanding of the ambitions of these multilateral actors. It is also important to spend time considering whether the Paris Agreement will change climate change policy significantly. This will give a good indication of where the multilateral climate change response is heading, and what this means in terms of a just response in the future. The focus on ambitions, achievements and future plans allows the chapter to provide a broad-brush overview of multilateral climate change governance and a comprehensive investigation into what has gone wrong, what has gone right and what is likely to change in future.

One possible explanation for the disconnect between the ambitions and achievements of multilateral actors is the ambiguous nature of the commitments made by multilateral actors in the Convention, Kyoto Protocol and Paris Agreement. The vague nature of these commitments is sometimes referred to as 'constructive ambiguity' because this ambiguity is not accidental (Stevenson and Dryzek 2014: 70). During negotiations, the Parties of the UNFCCC often seek flexible language to accommodate the diverging positions of parties. Common ambiguities include the use of 'shall' instead of 'will' or 'must', and the frequent use of 'should' instead of these more authoritative words (Stevenson and Dryzek 2014: 70). These ambiguities imply that commitments can be interpreted in different ways, which calls into question what the actors under the UNFCCC have committed to, and what this commitment will mean in practice. So, although it may appear that the commitment or ambitions of multilateral actors are in line with the demands of justice, this does not guarantee that the demands will be enabled. This will be further discussed throughout the chapter.

The chapter will now take each demand of justice in turn, and assess to what extent actors under the UNFCCC enable this demand. The assessment will make use of the four-point hierarchy developed in Chapter 4. Both the demands and the hierarchy are outlined below, as a reminder to the reader:

The Three Demands of Climate Justice

1. a) Global temperature changes must be kept at or below 2°C.
   b) Adaptation must be prioritised alongside mitigation.
2. The distribution of benefits and burdens in global climate change action must be based on the Polluter's Ability to Pay (PATP) model.

3. Capable actors, including individuals, firms, sub-state entities and international institutions, irrespective of the country in which they live or exist, must be held responsible for lowering emissions and/ or contributing to adaptation efforts, in line with their respective capabilities.

The Four-Point Hierarchy

1. Actors in the institution enable the demand of justice – the demand of justice is unequivocally fulfilled in its entirety.
2. Actors in the institution are consistently working towards enabling the demand of justice – the demand of justice is not yet fulfilled, but there are policies in place which are consistently leading towards this goal.
3. Actors in the institution have promised to begin working on enabling the demand of justice in the future – no policy has been adopted, but there is the potential for the creation of policy in order to consistently work towards enabling the demand of justice.
4. Actors in the institution do not enable the demand of justice – there has been no promise or attempt to enable the demand of justice and there are no policies in place.

DEMAND ONE – PROTECTING THE HUMAN RIGHT TO HEALTH

Demand One focuses on the protection of the human right to health and states that: a) global temperature changes must be kept at or below 2°C and b) adaptation must be prioritised alongside mitigation. Although keeping global temperatures at or below 2°C seems increasingly impossible due to current inaction, the IPCC maintains, at the time of writing, that there are multiple mitigation pathways that are likely to limit warming to below 2°C: emissions will have to be cut by 40%–70% by 2050 compared to 2010, and will need to be near zero or below in 2100 (IPCC 2014a: 14). The chapter will therefore assess to what extent the multilateral response to climate change is working towards these mitigation requirements.

Assessing whether adaptation is being prioritised alongside mitigation is slightly less straightforward. It has been estimated that global adaptation costs will be $125 billion USD in 2050 (Hof et al. 2010: 252). So, one aspect to examine is the extent to which the multilateral regime has been working towards raising these funds. However, adaptation is not merely a matter of financial cost, it is also about sharing technology and ensuring that countries vulnerable to climate change are assisted in preparing for climate change. As was discussed in Chapter 3, these aspects of adaptation are particularly important for the right to health because strengthening health systems through technology and knowledge transfer, rather than financial transfer alone, could significantly reduce the burden of disease (WHO 2013: 1). The discussion below will bear the complexities of adaptation in mind when evaluating to what extent multilateral actors are enabling the second part of Demand One.

Mitigation

Let us begin by examining the Convention. Chapter 3 explained that keeping global temperatures below 2°C serves mainly to protect future generations, since this threshold is not expected to be crossed until 2050–2100. It is therefore very encouraging that the Preamble of the Convention states that actors under the UNFCCC are 'determined to protect the climate system for present and future generations' (UNFCCC 1992: 6). This ambition is in line with Demand One, because if protecting future generations is seen as a reason to act on climate change, then this leaves space for emissions targets that enable this protection. The preamble sets the tone of the Convention, so this commitment is a good indication of the importance multilateral actors place on protecting future generations. However, since the preamble is not legally binding, it is important to investigate whether the remainder of the Convention reflects the sentiment of the preamble.

Encouragingly, the Convention highlights the importance of future generations in Article 3.1, which states that 'the Parties should protect the climate system for the benefit of present and future generations of humankind, on the basis of equity' (UNFCCC 1992: 9). The use of the word 'equity' is particularly interesting. Although equity does not imply equality, but rather fair treatment, this speaks to the idea that future generations must be protected, because they are mentioned as reasons to act against climate change without any conditions explaining that present generations are morally more important. This leaves space for policy on emissions reductions in protocols that follow the Convention. However, the wording of Article 3.1 is quite ambiguous. There is no guidance on how to balance the needs of present and future generations (Lawrence 2014: 100). This is a good example of a 'constructive ambiguity', or the ambiguity that is inserted into a commitment in order to accommodate the diverging positions of parties (Stevenson and Dryzek 2014: 70). These ambiguities are problematic, because they leave the question of what is actually being committed to very much open to interpretation. This may allow states to avoid action on climate change in future, because they can argue that their interpretation of obligations did not imply stringent action.

Although the ambiguous nature of Article 3.1 is indisputable, the wording of the Article nevertheless falls in line with the first part one of Demand One, because it represents a commitment to protect future generations alongside present generations. This is important, because enabling a demand of justice has been defined as creating a context under which this is possible. The commitments made in the Convention set the context for climate change action, in the sense that they set out the ambitions of the multilateral actors, allowing them to take action in line with these ambitions. Therefore, although the ambitions are, in the case of Article 3.1, ambiguous, they have the potential to create a context where the right to health of future generations is protected. Whether actors under the UNFCCC have managed to create this context is another matter that will be explored below.

First, it is important to point out that Article 1.1 of the Convention defines 'adverse effects to humankind' caused by climate change as effects on 'the operation of socio-economic systems or on human health and welfare' (UNFCCC 1992: 12). In addition, according to Article 4.1f, Parties agree to 'take climate change considerations into account . . . and employ appropriate methods . . . with a view to minimising adverse effects on the economy, on public health and on the quality of the environment' (UNFCCC 1992: 11). These Articles indicate that the Convention bases its concerns about climate change, at least partially, on human health. This is encouraging, because the mention of health creates a context under which the right to health can be protected. If multilateral actors acknowledge health as an important part of the climate change problem, then there is space to develop policies which protect health. Overall then, the Convention indicates that the actors under the UNFCCC are at least on the third rung on the hierarchy in terms of Demand One:

> 3. Actors in the institution have promised to begin working on enabling the demand of justice in the future – no policy has been adopted, but there is the potential for the creation of policy in order to consistently work towards enabling the demand of justice.

The potential to create policy lies in the ambitions of the Convention and the fact that a framework for the creation of policy has been established under the UNFCCC. Multilateral actors clearly have ambitions to protect future generations and to take human health into account, which indicates a potential for policy to make good on these ambitions. However, because the Convention does not set any mandatory mitigation targets and timetables, the ambitions in the Convention alone are not enough to fully protect the right to health. For this reason, it is important to investigate the Kyoto Protocol, to assess whether the multilateral actors live up to their ambitions and implement policies that enable the first part of Demand One. In international law, a Protocol can usually amend a treaty or add additional provisions. In the case of climate change, the Convention establishes a normative framework for developing mitigation and adaptation strategies, and the Kyoto Protocol contains specific provisions and regulations to achieve this. Therefore, ideally, the policies enshrined in the Kyoto Protocol should make good on the ambitions set out in the Convention.

Unfortunately, the Kyoto Protocol cannot be said to be seriously contributing to the protection of the human right to health for several reasons. First, the Kyoto Protocol did not originally include the 2°C target. This target was only affirmed at the 2009 Copenhagen Accord, and only applies to the second, and current, commitment period of the Kyoto Protocol. Second, the current commitment period of the Kyoto Protocol only aims to lower emissions by 'at least 18% below 1990 levels in the commitment period 2013 to 2020' (UNFCCC 2012: 4). The current IPCC report calls for emissions to be lowered by 40–70% of 2010 levels by 2050 to keep global average temperatures to a rise of 2°C. Lowering emissions to 18% of 1990 levels by 2020 is not in line with this requirement, because 1990 levels were substantially lower than 2010 levels. The growth rate of emissions increased from 1.5% a year in 1980–2000 to 3% a year in 2000–12 (J. Hansen et al. 2013: 1). The Kyoto Protocol's targets are therefore clearly out of line with what is required.

Third, the states held to account under the second period of the Kyoto Protocol only account for 15% of global GHG emissions (Lawrence 2014: 22). The distribution of responsibility between states under the Kyoto Protocol will be further discussed when assessing Demand Two, but this is a significant problem in terms of lowering global emissions. If global emissions are to be cut by 40–70% by 2050, it is woefully inadequate to only regulate 15% of emissions until 2020. Even if all the countries held to account under the Kyoto

Protocol halted all of their emissions immediately, this effort would be futile for the protection of the human right to health.

Fourth and finally, although the Kyoto Protocol sets specific targets, the Protocol does not set up any compliance mechanisms for failure to meet these targets. Instead, any emissions reductions which are not met can, 'on request . . . be added to the assigned amount for subsequent commitment periods' (UNFCCC 1998: 5). This is indicative of a very loose and voluntary compliance system, where states are not obligated to meet targets. This lack of compliance measures has proved problematic. Most countries signatory to the Kyoto Protocol are failing to meet the targets that have been set out (Stevenson and Dryzek 2014: 2). Furthermore, because countries do not face penalties for leaving the Kyoto Protocol, or for not joining in the first place, some countries, most notably the United States of America (USA), did not join the Protocol. As the USA is currently one of the top global emitters, and one of the richest countries in the world, this seriously undermines the potential of the Kyoto Protocol. In addition, some states, including Canada, Japan, Russia and New Zealand, have refused to participate in the second round of the Protocol when it became clear that they could not fulfil their commitments. These are major emitters, and their lack of participation raises grave concerns over whether actors under multilateral governance enable Demand One of justice.

From the above, it is clear that multilateral actors are not creating a context where Demand One can be met, because emissions are not being regulated in a way which protects the human right to health. In fact, global emissions have only recently begun to slow down, growing by 0.7% in 2014, stalling in 2015 and growing by 0.2% in 2016 (Carbon Brief 2016).[1] Under Kyoto Protocol policies, global temperatures are set warm by between 2.7°C and 4.9°C, reaching 3.6°C by 2100 (Climate Action Tracker 2016). It is very clear that, so far, the multilateral impact on the global level of emissions has not been in line with what is necessary to achieve the goal of no more than 2°C warming. For this reason, actors under the UNFCCC continue to sit on the third rung of the four-point hierarchy in terms of the first part of Demand One:

> 3. Actors in the institution have promised to begin working on enabling the demand of justice in the future – no policy has been adopted, but there is the potential for the creation of policy in order to consistently work towards enabling the demand of justice.

The potential to create policy lies in the ambitions of the Convention and the fact that the Kyoto Protocol has created policies which, although not in line with the targets suggested by the IPCC, nonetheless imply that there is an existing policy framework that can be built upon. This leads the chapter to its assessment of the Paris Agreement. If multilateral actors are so far failing to enable the first part of Demand One, it is important to investigate whether this is likely to change in the future. At first glance, there are some encouraging signs that it might, because the Paris Agreement explicitly mentions the right to health as a reason to act on climate change in its preamble (UNFCCC 2015: 21):

> Acknowledging that climate change is a common concern of humankind, Parties should, when taking action to address climate change, respect, promote and consider their respective obligations on human rights, **the right to health**, the rights of indigenous peoples, local communities, migrants, children, persons with disabilities and people in vulnerable situations and the right to development, as well as gender equality, empowerment of women and intergenerational equity.

This is an interesting development, because although the Convention mentions health, the 'right to health' was not included in the Convention or Kyoto Protocol. In fact, the Paris Agreement is the first multilateral environmental agreement to recognise human rights (ENB 2015c: 43). This is a significant step forward. If the Paris Agreement unambiguously aims to protect the human right to health, this opens up the space to create policy that does so. The preamble of the Agreement also makes mention of international equity, which was alluded to in the Convention under Article 3.1. This direct reference to international equity is important, because although the concept of equity is left ambiguous, this allows room for protecting future generations. However, under international law, the preamble of a treaty is not considered legally binding. For this reason, it is important to investigate how exactly the UNFCCC aims to protect the right to health under the Paris Agreement. This will be more telling than merely setting out the ambition to do so.

Encouragingly, the Paris Agreement has been widely commended for specifying global temperature targets which are stricter than any previous targets set out by the UNFCCC. Multilateral actors have, for the first time in multilateral climate change governance history, set a target that is below 2°C. Article 2 of the Paris Agreement specifies the commitment to limit 'the increase in the global average temperature to well below 2°C above pre-industrial levels and to pursue efforts to limit the temperature increase to 1.5°C above pre-industrial levels' (UNFCCC 2015: 22). The celebration around this new target is not entirely unfounded, because it is quite clearly a positive step forward in terms of protecting the human right to health, which requires that temperatures are held to below 2°C. However, although this ambitious aim is a positive development in terms of creating a context where the human right to health can be protected, there have been some important changes in how emissions targets are formulated. The Kyoto Protocol set out globally agreed emissions targets and countries were assigned obligations based on this target. This 'top-down' model has now been replaced. Instead, under the Paris Agreement, countries are expected to submit their own 'Intended Nationally Determined Contributions', or INDCs.

The INDC model means that the Paris Agreement sets no long-term global mitigation timeline, instead leaving decisions on mitigation up to individual Parties. In a sense, the multilateral climate regime is evolving from a 'global deal' model, in which countries negotiate emissions targets, to a 'pledge-and-review' model, in which each country defines its own goals, subject to some form of intergovernmental review (Sander et al. 2015: 469). This is potentially problematic for the human right to health, because if each Party can determine what goes into their contribution, there is no guarantee that the INDCs will collectively result in emission reductions that protect this right. In fact, the current INDCs are not in line with what is required to meet the 1.5°C target set out in the Paris Agreement. Only five out of 185 INDCs are rated as 'sufficient' enough to hold global temperatures below 2°C (Climate Action Tracker 2016a). These 'sufficient' pledges only represent 0.4% of global emissions. If INDCs remain unchanged, temperatures will be kept between 2.2°C and 3.4°C, with a median warming of 2.7°C by 2100 (Climate Action Tracker 2016a). In this sense, the current INDCs are clearly not consistent with what has been agreed at Paris, and furthermore are not adequate in terms of protecting the human right to health. It remains to be seen if these INDCs are adequately revised before the Paris Agreement comes into force.

In addition, it is not clear whether the Paris Agreement will fare better than the Kyoto Protocol in terms of holding Parties to account for their emissions reductions plans. The Paris Agreement sets out several measures that could encourage compliance. For example, Article 4 calls for INDCs to be updated every five years, with each successive pledge required to be as stringent as, or more ambitious than, the existing one (UNFCCC 2015: 23). In addition to this, the Paris Agreement calls for a 'global stocktake' under Article 14, beginning in 2023 and scheduled to occur every five years thereafter. The results of this stocktake 'shall inform Parties in updating and enhancing, in a nationally determined manner, their actions . . . as well as in enhancing international cooperation for climate action' (UNFCCC 2015: 29). This global stocktake, combined with a review of INDCs every five years, has the potential to create a context where compliance is ensured, because the pressure of monitoring progress may discourage stalling of mitigation efforts. In addition, the Paris Agreement establishes a compliance mechanism in Article 15, which will consist of a committee of experts. However, this committee will 'function in a manner that is transparent, non-adversarial and non-punitive' (UNFCCC 2015: 29). The term 'non-punitive' suggests that the compliance mechanism will have a soft touch, and that much like in the Kyoto Protocol, there are currently no established consequences for non-compliance.

Indeed, participation in the Paris Agreement is entirely voluntary and states can choose to withdraw at any time after the first three years of the implementation of the Paris Agreement under Article 28 (UNFCCC 2015: 32). The current president of the United States, Donald Trump, plans to do just that. It remains to be seen whether other countries follow suit. Of course, voluntary membership of the Paris Agreement comes with advantages: the voluntary nature of participation allows 'for higher ambition in [states'] mitigation and adaptation actions' (UNFCCC 2015: 24) Nevertheless, weak compliance measures have been quite problematic under the Kyoto Protocol, as was discussed above. The Paris Agreement has not adequately moved forward on this issue, and it is questionable whether states will comply with the targets they have set. Historically, most governments have over-pledged and under-delivered on their climate change action plans (Harris and Lee 2017: 13), which does not raise great hopes for the new 'bottom up' regime, not least in terms of protecting the human right to health.

However, although the Paris Agreement still faces significant challenges, it is nevertheless promising that multilateral actors continue to work together, and have created a second binding treaty. This is indicative of an emerging context within which the right to health of future generations can be protected. The mention of the right to health and movement towards increased mitigation efforts are especially important in terms of creating this context. Multilateral actors continue to move forward, if slowly. Of course, it is difficult to predict whether the Paris Agreement will protect the human right to health once it comes into force. This will depend on whether INDCs are revised, and whether countries live up to their promises and comply with the global temperature targets that have been set. These remaining problems will be further discussed in the final part of the chapter. For now, the chapter turns its attention to part two of Demand One.

Adaptation

The second part of Demand One calls for the prioritisation of adaptation alongside mitigation. Assessing to what extent adaptation has been prioritised is not a straightforward process. One important area of focus is the extent to which the multilateral regime has been working towards raising funds for adaptation. However, adaptation is not merely a matter of financial cost, it is also about sharing technology and ensuring that countries vulnerable to climate change are assisted in preparing for climate change. The assessment will therefore focus on finance, technology and assistance. In this way, the chapter endeavours to take the complexities of adaptation into account. Of course, not all complexities can be captured in this short discussion. Nevertheless, by assessing the main promises made and policies created around adaptation, the chapter provides an overview of multilateral action on adaptation that can be built upon in future research.

Let us begin by examining the Convention. The first mention of adaptation is in Article 4.1e, which calls on all parties to 'cooperate in preparing for adaptation to the impacts of climate change; develop and elaborate appropriate and integrated plans for coastal zone management, water resources and agriculture, and for the protection and rehabilitation of areas, particularly in Africa, affected by drought and desertification, as well as floods' (UNFCCC 1992: 11). This is quite comprehensive, in the sense that it outlines specific policy areas to focus on and makes reference to vulnerable regions. The comprehensive nature of the first mention of adaptation is promising, because setting out what is required allows for the creation of a context where adaptation can be pursued. Furthermore, because adaptation is not qualified as secondary or less important than mitigation, this creates a context where adaptation could be prioritised alongside mitigation.

Encouragingly, the Convention does not merely outline the need for adaptation. It goes further, by setting out two responsibilities to enable adaptation: financial and technological transfer. Article 4.4 states that 'the developed country Parties ... shall assist the developing country Parties that are particularly vulnerable to the adverse effects of climate change in meeting costs of adaptation to those adverse effects' (UNFCCC 1992: 11). Furthermore, Article 4.5 states that 'the developed country Parties ... shall take all practicable steps to promote, facilitate and finance, as appropriate, the transfer of, or access to, environmentally sound technologies and know-how to other Parties, particularly developing country Parties' (UNFCCC 1992: 14). Setting out responsibilities for both financial and technological transfer is promising, because this creates a context where adaptation can be prioritised alongside mitigation. If specific responsibilities are set out, it is clear what is expected in order to act on adaptation needs. This makes it possible for individual states to develop policy on adaptation, and for multilateral actors to create global policies that ensure adaptation is prioritised.

In addition to setting out responsibilities, the Convention also outlines the need for a financial mechanism 'for the provision of financial resources on a grant or concessional basis, including for the transfer of technology' in Article 11 (UNFCCC 1997: 22). Although the Convention does not set out more than a few initial plans for this mechanism, it is clear from Article 11.4 that Parties are expected to work out the finer details of this mechanism as soon as possible: parties must 'implement the above-mentioned provisions at [the] first session [COP1, Berlin 1995]' (UNFCCC 1992: 23). In addition, Article 11.4 stipulates that 'within four years thereafter, the Conference of the Parties shall review the financial mechanism and take appropriate measures' (UNFCCC 1992: 23). In this sense, the Convention not only sets out plans for the creation of a financial mechanism, but also puts procedures for updating this mechanism in place. This is important in terms of creating a context where adaptation can be prioritised alongside mitigation, because it sets out clear measures that can make this possible.

The final way in which the Convention helps to create a context where adaptation can be prioritised alongside mitigation is by setting out a list of countries that are particularly vulnerable to climate change effects and must therefore be prioritised in Article 4.8 (UNFCCC 1992: 15):

a) Small island countries;
b) Countries with low-lying coastal areas;
c) Countries with arid and semi-arid areas, forested areas and areas liable to forest decay;
d) Countries with areas prone to natural disasters;
e) Countries with areas liable to drought and desertification;
f) Countries with areas of high urban atmospheric pollution;
g) Countries with areas with fragile ecosystems, including mountainous ecosystems;
h) Countries whose economies are highly dependent on income generated from the production, processing and export, and/or on consumption of fossil fuels and associated energy-intensive products; and
i) Land-locked and transit countries.

By listing these countries, the Convention further clarifies what adaptation will require, and who must be prioritised. There is a clear idea of who the most vulnerable are, and why they must be protected. In addition to this, the Convention defines adaptation as important, allocates responsibilities for adaptation and sets out specific mechanisms to enable adaptation. Overall, then, multilateral actors have set out clear ambitions to act on adaptation, or in other words promised to act on adaptation in future. For this reason, multilateral actors fall on the third rung of the four-point hierarchy in the case of the second part of Demand One:

> 3. Actors in the institution have promised to begin working on enabling the demand of justice in the future – no policy has been adopted, but there is the potential for the creation of policy in order to consistently work towards enabling the demand of justice.

The potential to create policy lies in the ambitions of the Convention and the fact that a framework for the creation of policy has been established under the UNFCCC. In order to examine whether such policies have indeed been put into place, let us now turn to the Kyoto Protocol. Encouragingly, the Kyoto Protocol sets out policies on adaptation, which indicates that it is being prioritised alongside mitigation. Article 3.14 of the Protocol, for example, states that the Parties of the Protocol must 'consider what actions are necessary to minimise the adverse effects of climate change and/or the impacts of response measures on developing country Parties. Among the issues to be considered shall be the establishment of funding, insurance and transfer of technology' (UNFCCC 1998: 5). This relates directly back to the responsibilities for funding and transfer of technology set out in the Convention, which implies that by agreeing to the Kyoto Protocol, multilateral actors are attempting to institutionalise the adaptation-related ambitions of the Convention. Indeed, the Kyoto Protocol establishes several ways in which funding and technological transfer for adaptation can be achieved.

First, the Kyoto Protocol sets up three specific funds to assist less developed countries: the Special Climate Change Fund, the Least Developed Countries Fund and the Adaptation Fund. The fact that the Kyoto Protocol sets up these climate funds indicates that actors under the UNFCCC are setting out policies to enable the second part of Demand One. Specific funds for adaptation help to create a context where adaptation can be prioritised, because multilateral actors can contribute to these funds, therefore making adaptation possible in the first place. However, although the Kyoto Protocol has put policies into place, it is questionable whether these are consistently leading toward a context where the second part of Demand One can be met (which would place multilateral actors on rung two). The volume of the funds set up by the Kyoto Protocol is very small compared to the anticipated cost of adaptation in developing countries (Shrivastava and Goel 2010: 120).

It is estimated that global adaptation costs will be $125 billion USD in 2050. At the time of writing, the three funds named above hold a total of $1.7 billion USD, collectively. The Adaptation Fund has pledges of $483 million USD, the Least Developed Countries Fund has pledges of $914 million USD and the Special Climate Change Fund has pledges of $347 million USD. $1.7 USD in 2017 does not bode well for raising $125 billion USD by the time it is required in 2050. Not only are current funds inadequate, less developed countries have complained that the complexity of current arrangements significantly constrains their access to funds for adaptation project activities (Möhner and Klein). This does not create a context where adaptation can be prioritised, because an important part of adaptation is covering costs as quickly as possible. On the continent of

Africa alone, the annual adaptation costs grow by 10% each year under current emissions levels (Abeysinghe and Huq 2016: 195). Funds must become available, and the sooner the better, to prevent spiralling costs and significant impacts on human health for the most vulnerable countries.

There has been more optimism surrounding the new 'Green Climate Fund' (GCF), established at Durban in 2011, when the second commitment period of the Kyoto Protocol was agreed to. The GCF is expected to make a significant and ambitious contribution to the global adaptation efforts (Vanderheiden 2015: 34; Lawrence 2014: 107). Funds will go towards several different areas: incremental costs for activities relating to adaptation, mitigation, technology development and transfer, and capacity building. The GCF is anticipated not only to complement, but perhaps eventually to replace, other multilateral climate change funds discussed above (Vanderheiden 2015: 34). At the time of writing, the GCF has received $10.3 billion USD in pledges by 33 countries (Climate Funds Update 2017). Although this is promising, because the Fund aimed to receive $10 billion USD by the end of 2014 and is therefore largely on target, not all countries are committed to assisting the GCF (Paddy 2014). The previous prime minister of Australia, Tony Abbott, notably refused to contribute, claiming that the GCF is 'socialism masquerading as environmentalism' (Goldenberg 2014). Furthermore, although former US president Barack Obama deposited $500 million USD to the GCF before the end of his final term, Donald Trump has made no further payments, calling the GCF 'yet another scheme to redistribute wealth out of the United States' (Kotchen 2017). With the GCF dependent on national leadership decisions, it is questionable whether enough funds for adaptation will be raised in time. It is not only the GCF that faces these problems. All UNFCCC funds established for supporting adaptation are voluntary in nature and are not linked to any formal compliance measures (Abeysinghe and Huq 2016: 196). In this sense, then, although policy has been created, multilateral actors are struggling to create a context where adaptation is prioritised alongside mitigation.

This is especially true given that less developed countries have regularly lamented the lack of adequate, predictable and long-term climate finance, and accuse developed countries of reneging on their promises (Chukwumerije and Coventry 2016: 837). Developed countries, however, insist they are doing their best in very tough economic conditions and express concern that some less developed countries are attempting to use climate change as an excuse to get developed countries to fund their national economic development (Chukwumerije and Coventry 2016: 837). These fundamental disagreements are hampering action on adaptation. Furthermore, although developed countries may have agreed to technological transfer in Article 3.14, there was disagreement over what this requires when the Kyoto Protocol was negotiated. Developed countries emphasised the transfer of technical information, while less developed countries emphasised technology transfer on non-commercial and preferential terms as most important (ENB 1997: 4). So, although the Kyoto Protocol played an important role in policy creation, the ambiguous language surrounding technology transfer suggests that the extent to which technology is in fact shared with less developed countries remains to be seen.

From all of the above, it is perhaps unsurprising that scholars and policymakers alike claim that adaptation has historically been seen as a marginal policy option – mitigation's 'poor cousin' in the climate policy arena (Ayers et al. 2010: 270). Because the ultimate objective of the Convention was set as stabilising GHGs in the atmosphere, the adaptation agenda only emerged in the shadow of mitigation during the UNFCCC negotiations (Ayers et al. 2010: 196). It has been suggested that adaptation has not been prioritised alongside mitigation because the performance of adaptation options cannot be measured and expressed in a single metric, for example carbon (Klein and Persson 2008: 1). This renders it difficult for decision-makers to compare alternative adaptation options and to consider potential trade-offs (Klein and Persson 2008: 1). However, it is clear that even in the face of these difficulties, changes need to be made to ensure that adaptation is prioritised and the human right to health is protected. The current multilateral climate change response, under the Kyoto Protocol, is not conducive for fair and effective action on adaptation for many less developed countries (Ayers et al. 2010: 271). This indicates that a context in which the second part of Demand One can be met has so far not been created by multilateral actors. For this reason, actors under the UNFCCC remain on the third rung of the four-point hierarchy:

> 3. Actors in the institution have promised to begin working on enabling the demand of justice in the future – no policy has been adopted, but there is the potential for the creation of policy in order to consistently work towards enabling the demand of justice.

The potential for the creation of policy lies in the ambitions of the Convention and the fact that the Kyoto Protocol incorporates policies that, although inadequate, are attempting to meet these ambitions. Nevertheless, the failure to ensure financial and technological transfer presents a key hindrance that must be overcome in future in order to bring about a more just response to climate change. To examine to what extent multilateral actors have moved forward on these matters since the Kyoto Protocol, the chapter now turns to assessing the run-up to the Paris Agreement and the Paris Agreement itself.

One key development has been the establishment of the National Adaptation Programmes of Action (NAPAs), set out at the 2001 COP7 in Marrakesh. NAPAs provide a process for Least Developed Countries to identify priority activities that respond to their urgent and immediate needs to adapt to climate change. In order to assist countries with preparation of their NAPAs, multilateral actors also set up a Least Developed Country Expert Group (LEG) at COP7. The LEG can provide guidance and advice on the preparation and implementation strategy for NAPAs, as well as providing technical guidance and advice on the implementation and revision of NAPAs. In addition to assistance from the LEG, Least Developed Countries can also apply for funding for NAPA preparation, implementation and revision through the Least Developing Country Fund. Every Least Developed Country has now submitted a NAPA, allowing multilateral actors to understand exactly what is required for adaptation in the most vulnerable countries. The first NAPA was submitted by Samoa in December 2005, with the most recent one being submitted by South Sudan in November 2016. The fact that every Least Developed Country has now submitted a NAPA is an excellent step in the right direction.

NAPAs are an important element of prioritising mitigation alongside mitigation. If countries are able to report on what they need and have access to financial mechanisms as well as technical and policy advice, this makes it possible for adaptation to be acted on. Unfortunately, despite every country having submitted a NAPA, many countries are still waiting to implement theirs (Abeysinghe and Huq 2016: 196). The main barrier for implementation has been insufficient financial resources. The Copenhagen COP15 in 2009 generated important political momentum in climate finance, with developed countries pledging a 'fast-start' of $30 billion USD in 2010–12 and reaching $100 billion USD a year by 2020, with a balanced allocation between adaptation and mitigation (Chukwumerije and Coventry 2016: 843). However, only around one fifth of finances have so far been allocated to adaptation (Abeysinghe and Huq 2016: 197). In this sense, although policy has been established, multilateral actors continue to struggle to create a context where adaptation is prioritised alongside mitigation.

A second important step in adaptation policy was the 2010 COP16 Cancun Adaptation Framework, which set out two important mechanisms. The first is National Adaptation Plans (NAPs), which follow on from NAPAs. Instead of focusing on immediate concerns, NAPs focus on medium- and long-term adaptation needs of both least developed countries and developing countries more broadly. This is important, because many adaptation needs are not immediate, but must be addressed in the long term to protect the human right to health. The second mechanism set out by the Cancun Adaptation Framework is the Adaptation Committee, which aims to promote the implementation of enhanced action on adaptation. Members of the committee provide technical support and guidance, and share relevant knowledge, experience and good practices. This is an important part of prioritising adaptation, because knowledge around adaptation is crucial for effective and far-reaching implementation. In addition, the Adaptation Committee provides a place for countries to communicate their monitoring and review of adaptation actions, as well as the support they have received. This allows multilateral actors to keep records of what help has been received, and whether this has been adequate. Such knowledge is important for informing policymakers about what must be prioritised. Finally, the Adaptation Committee provides information and recommendations at the COPs to facilitate decision-making on finance, technology and capacity building. This is a key part of prioritising adaptation, because it allows policymakers to negotiate adaptation and make well-informed decisions on what should be prioritised. All in all, then, there has been significant progress in the run-up to the Paris Agreement.

Let us now turn to assessing the Paris Agreement itself to get a sense of what lies ahead for multilateral climate change policy. One important decision made at the COP21 negotiations in Paris was that INDCs could include adaptation plans and set out funding and technology needs. This is something that less developed countries pushed for, while developed countries wanted the INDCs to be focused on mitigation (Chukwumerije and Coventry 2016: 842). Having INDCs include adaptation plans and needs is a significant victory, because it allows less developed countries to ensure that developed countries are aware of their adaptation needs outside of the NAPAs, which have yet to be meaningfully acted upon. Other notable achievements of the Paris Agreement are that it sets out several specific adaptation goals under Article 7, as well as goals for technology and financial transfer under Articles 9 and 10.

Article 7.1 establishes a 'global goal on adaptation of enhancing adaptive capacity, strengthening resilience and reducing vulnerability to climate change, with a view to contributing to sustainable development and ensuring an adequate adaptation response in the context of the temperature goal referred to in Article 2' (UNFCCC 2015: 25). In addition, Article 7.6 notes the 'importance of support for and international cooperation on adaptation efforts and the importance of taking into account the needs of developing country Parties, especially those that are particularly vulnerable to the adverse effects of climate change' (UNFCCC 2015: 25). By stressing the importance of adaptation, multilateral actors create a context where this is prioritised alongside mitigation. Furthermore, Article 7.7 of the Paris Agreement calls on countries to share 'information, good practices, experiences and lessons learned' and to strengthen 'institutional arrangements [. . .] to support the synthesis of relevant information and knowledge, and the provision of technical support and guidance' (UNFCCC 2015: 25).

In addition, Parties are expected to assist 'developing country Parties in identifying effective adaptation practices, adaptation needs, priorities, support provided and received for adaptation actions and efforts, and challenges and gaps, in a manner consistent with encouraging good practices', with the goal of 'improving the effectiveness and durability of adaptation actions' (UNFCCC 2015: 25). Here, the Paris Agreement is setting out clear goals and expectations, which helps to create a context where adaptation is taken seriously and prioritised alongside mitigation. Finally, Article 7.14 notes that the global stocktake will not only review mitigation targets, but also 'review the adequacy and effectiveness of adaptation and support provided for adaptation; and review the overall progress made in achieving the global goal on adaptation' (UNFCCC 2015: 26). Again, this is very encouraging, because it allows policymakers to ensure that adaptation measures are indeed being acted upon, something the multilateral regime has struggled with in the past.

Article 9.1 calls on 'developed country Parties [to] provide financial resources to assist developing country Parties with respect to both mitigation and adaptation' (UNFCCC 2015: 26). In addition, Article 9.9 sets out goals to ensure that adaptation financing is approved: Parties 'shall aim to ensure efficient access to financial resources through simplified approval procedures and enhanced readiness support for developing country Parties, in particular for the least developed countries and small island developing States, in the context of their national climate strategies and plans' (UNFCCC 2015: 26). These goals are encouraging, placing responsibility directly on developed countries, and focusing on the lack of financial transfer that has been an obstacle in achieving adaptation in the past. Similarly, Article 10.1 states that 'Parties share a long-term vision on the importance of fully realising technology development and transfer in order to improve resilience to climate change and to reduce greenhouse gas emissions' (UNFCCC 2015: 26). Furthermore, Article 10.4 establishes a 'technology framework' in order to promote and facilitate 'enhanced action on technology development and transfer'. These Articles appear quite promising at first glance, with multilateral actors seemingly prioritising adaptation by establishing a new mechanism for technology transfer and being very clear on the responsibility of developed countries.

However, although finance and technology transfer are explicitly discussed in Articles 9 and 10, the Paris Agreement does little to ensure funding requirements will be met or that vulnerable countries will be able to design and implement measures to meet their adaptation needs (Chukwumerije and Coventry 2016: 842). For one, no new financial target was stipulated in the Paris Agreement. It is also unclear how financing will be measured and therefore monitored to ensure developed countries are meeting the commitments they have made (Chukwumerije and Coventry 2016: 843). No countries were assigned a specific amount they must donate, and there are no mandatory targets (Santos 2017: 14). In addition, although the Paris Agreement set out a new technology mechanism, the focus is mainly on innovation and enabling development of technology, with passing reference to removing barriers that prevent the transfer of existing technology (Chukwumerije and Coventry 2016: 845). In this sense, it seems that multilateral actors continue to struggle to prioritise adaptation adequately, which puts the human right to health at serious risk.

Nevertheless, the multilateral regime has made good progress in terms of monitoring adaptation needs through NAPAs, NPAs and INDCs, and the Paris Agreement appears to move forward on adaptation policy, at least in terms of what has been promised. The creation of the LEG and Adaptation Committee are also encouraging. And yet, multilateral actors continue to struggle to ensure adequate financial and technological transfer. Whether this is likely to change in future will be further discussed in the final part of the chapter. For now, the chapter turns to Demand Two.

DEMAND TWO – THE RESPONSIBILITIES OF STATES

Demand Two states that the distribution of benefits and burdens in global climate change action must be based on the Polluter's Ability to Pay (PATP) model.[2] In this sense, enabling Demand Two requires that the responsibility to contribute finances and lower emissions is based on both emissions levels and wealth levels. The chapter will investigate to what extent multilateral actors are enabling this model of responsibility by examining the Convention, Kyoto Protocol and Paris Agreement in turn. The discussion below will also consider to what extent less developed countries have been part of decision-making on climate responsibility. Chapter 3 explained that these actors are often left out of decision-making procedures, and the allocation of climate responsibilities is a prime example of this. The chapter will therefore not only discuss the content of the Convention, Kyoto Protocol and Paris Agreement, but also examine how these documents came to be written and assess to what extent voices were left out or ignored in the process.

In order to assess how responsibility has been distributed, the chapter will examine the notion of 'common but differentiated responsibility' (CBDR). The concept of CBDR has determined how much each country must contribute to the climate change problem for decades. During this time, there have been intense disputes over how exactly to interpret, or define, CBDR (Chukwumerije and Coventry 2016: 837). To this day, CBDR has different meanings in different parts of the world – with less developed countries stressing the importance of past emissions, and developed countries preferring a definition that focuses on capability to act on climate change. To understand where this difference in interpretation stems from, the chapter will explore how CBDR was defined in the Convention and how it was interpreted when states came to negotiate the Kyoto Protocol and the Paris Agreement. This will allow the chapter to explore whether multilateral actors have enabled a PATP model of responsibility through the CBDR, and also examine to what extent less developed countries have been included in decision-making on climate responsibilities.

The first mention of the CBDR can be found in the preamble of the Convention, which acknowledges that 'the global nature of climate change calls for the widest possible cooperation by all countries and their participation in an effective and appropriate international response, in accordance with their common but differentiated responsibilities and respective capabilities and their social and economic conditions' (UNFCCC 1992: 2). At first glance, this definition of CBDR seems very ambiguous, because it does not specify who has what responsibilities and capabilities – it merely explains that they are both important factors. However, a few pages later, the preamble of the Convention states that there is a 'need for developed countries to take immediate action in a flexible manner on the basis of clear priorities, as a first step towards comprehensive response strategies at the global, national and, where agreed, regional levels that take into account all greenhouse gases, with due consideration of their relative contributions to the enhancement of the greenhouse effect' (UNFCCC 1992: 4). This second mention of CBDR directly states that responsibility should, in the first instance, be allocated according to emissions levels. The phrase 'due consideration of their relative contributions to the enhancement of the greenhouse effect', in particular, indicates a polluter pays (PPP) based approach, rather than an Ability to Pay (ATP) or Polluter's Ability to Pay (PATP) approach.[3] However, the preamble of the Convention is not legally binding. It is therefore important to examine the main body of the Convention.

Article 3.1 of the Convention states that 'the Parties should protect the climate system [. . .] in accordance with their common but differentiated responsibilities and respective capabilities. Accordingly, the developed country Parties should take the lead in combating climate change and the adverse effects thereof' (UNFCCC 1992: 9). This definition of the CBDR echoes the preamble because it assigns responsibility to developed countries in the first instance, but it is much more ambiguous in its wording, particularly in terms of defining capability. This lack of discussion of capability is another good example of a 'constructive ambiguity' which is put in place to accommodate the diverging positions of parties (Stevenson and Dryzek 2014: 70). By not defining capability, the definition is left open to interpretation.

This is different to the preamble, where contributions to the greenhouse effect are explicitly mentioned as reasons to act. The ambiguity was left in place because developed countries were hesitant to have their emissions levels determine how much they should contribute.

Although they agreed to taking greater responsibility for climate change, as defined in Article 3.1, developed countries wanted their responsibility to be assigned because of their greater capacity to pay, not because of their historical responsibility (Bodansky 1993: 480). In this sense, developed countries advocated for an ATP, rather than a PPP model of responsibility. This differs from what the less developed countries had hoped for when agreeing to CBDR, which was that developed countries would be held to account because they are the predominant cause of climate change (Bodansky 1993: 479), or in other words, a PPP model of responsibility.

These differences in interpretation are quite revealing of how the negotiations played out. It is the developed country ATP model of CBDR which is written into the legally binding part of the Convention, while the less developed country PPP model of CBDR is only mentioned in the non-legally binding preamble. This is indicative of less developed countries' voices being ignored, or at least downplayed, when the Convention was written. Not including less developed country demands in the legally binding parts of the Convention is perhaps unsurprising. Multilateral climate change actors have consistently been accused of excluding less developed countries, and particularly Least Developed Countries, from important decision-making procedures. Their exclusion is a result of a multitude of factors. For one, less developed countries typically have much smaller delegations and therefore cannot keep abreast of the myriad issues, meetings and side-events at the COPs (Eckersley 2012: 35). During a typical two-week COP, dozens of meetings run in parallel. Although simultaneous interpretation is provided in the six languages of the United Nations during formal plenary meetings, during informal consultations and working group meetings negotiators are left to rely on their own English skills for communication (Stevenson and Dryzek 2014: 74). It is at these informal meetings where most actual negotiation is known to take place (Stevenson and Dryzek 2014: 74). Vanderheiden (2008: 253) goes so far as to claim that less developed countries are usually offered a 'take it or leave it' deal after the United States and European Union have made decisions without them. Similarly, Bulkeley and Newell (2010: 31) argue that less developed countries have little capacity to shape or influence climate governance.

As will be seen below, the initial decision on how to interpret CBDR has influenced decisions ever since. How this has played out in terms of policy will be discussed below, when assessing the Kyoto Protocol and Paris Agreement. First, it should be noted that in terms of enabling Demand Two, the ambiguity and difference in opinion on CBDR does not necessarily imply that the PATP cannot be enabled. The Convention outlines both a PPP and ATP notion of responsibility, in the preamble and main body of the Convention, respectively. If both definitions are considered equally important, this would mean the CBDR is in line with the PATP model advocated in Chapter 3, which calls on states to be held to account for climate change action according to their emissions and wealth levels. So, even though the Convention does not directly promote a PATP model, it seems that there may be some flexibility to accommodate this model, because the Convention mentions both historical emissions and capabilities as a reason for developed countries to be the first to act on climate change. This is important, because this flexibility allows for room for the creation of a context where Demand Two can be met. For this reason, actors under the UNFCCC sit on the third rung in terms on enabling Demand Two:

3. Actors in the institution have promised to begin working on enabling the demand of justice in the future – no policy has been adopted, but there is the potential for the creation of policy in order to consistently work towards enabling the demand of justice.

The potential to create policy lies in the flexibility of the CBDR and the fact that a framework that can determine the allocation of climate responsibility was established under the UNFCCC in 1992. To examine how this potential for policy has played out, the chapter now turns to examining the Kyoto Protocol. As was explained above, ideally, the Kyoto Protocol should make good on the ambitions set out in the Convention in the form of policy, and therefore create a context where the PATP is enabled. At first glance, it seems like this is the case. The preamble of the Kyoto Protocol stipulates that the Protocol is 'guided by Article 3' of the Convention, which outlines the CBDR (UNFCCC 1998: 81). In addition, CBDR is explicitly reaffirmed in Article 10 of the Kyoto Protocol, which states that 'all Parties, taking into account their common but differentiated responsibilities' shall act on climate change (UNFCCC 1998: 9). This seems to indicate that multilateral actors aim to implement the CBDR as part of the Kyoto Protocol, leaving room, or scope, for the enablement of the PATP.

However, this scope is restricted, because being guided by Article 3 of the Convention implies an ATP approach to responsibility, rather than a PATP approach. Furthermore, the ATP approach is the preferred approach of developed countries, indicating that less developed countries were not able to include their PPP interpretation of CBDR in the Kyoto Protocol. This is perhaps unsurprising, given that the Kyoto protocol was written only three years after the Convention, which did not give less developed countries much time to increase their capacity to negotiate. However, this is another clear example of less developed countries not being included in decision-making procedures to the same extent as developed countries. As will be seen below, defining CBDR in line with the ATP rather than PPP model has continued to persist in the run-up to the Paris Agreement. Before this can be discussed, the chapter must examine to what extent the Kyoto Protocol enables the PATP model, if at all. More specifically, it is important to examine whether the highest emitters and wealthiest nations are held to account.

Unfortunately, the Kyoto Protocol leaves out both high emitters and wealthy nations from the scope of responsibility. Article 2.2 of the Koto Protocol states that 'the Parties included in Annex I shall pursue limitation or reduction of emissions of greenhouse gases' (UNFCCC 1998: 2), and Article 11 states that Parties included in Annex II shall: '(a) Provide new and additional financial resources (b) including for the transfer of technology' (UNFCCC 1998: 10–12). In other words, Annex I countries are held to account for emissions reductions, and Annex II countries, which represent the OECD members of Annex I, are held to account for financial contributions under the Kyoto Protocol. These countries are listed below:

Annex I – Australia, Austria, Belarus, Belgium, Bulgaria, Canada, Czechoslovakia, Denmark, European Economic Community, Estonia, Finland, France, Germany, Greece, Hungary, Iceland, Ireland, Italy, Japan, Latvia, Lithuania, Luxembourg, Netherlands, New Zealand, Norway, Poland, Portugal, Romania, Russian Federation, Spain, Sweden, Switzerland, Turkey, Ukraine, United Kingdom of Great Britain and Northern Ireland, United States of America.

Annex II – Australia, Austria, Belgium, Canada, Denmark, European Economic Community, Finland, France, Germany, Greece, Iceland, Ireland, Italy, Japan, Luxembourg, Netherlands, New Zealand, Norway, Portugal, Spain, Sweden, Switzerland, Turkey, United Kingdom of Great Britain and Northern Ireland, United States of America.

Non-Annex I countries, a category that includes 148 countries including Brazil, China, India and South Africa, are considered 'mostly developing countries' and are not held to account for emissions reductions or financial contributions under the Kyoto Protocol. Examining the list of Annex I countries responsible for lowering emissions above, it quickly becomes evident that many of the highest-emitting countries are not held to account for lowering emissions under the Kyoto Protocol. India is currently in third place for global emissions shares, and China is the world's largest emitter in absolute terms, ranking on a par with the EU in per capita terms. It would of course be unfair to criticise multilateral actors for not holding China and India to account in 1997 when the Protocol was first written, because these countries did not have high emissions levels at the time. However, the Kyoto Protocol is now in its second commitment period, which is in effect until 2020, and these countries are still not held to account for emissions reductions under the Kyoto Protocol. Furthermore, the USA, which is, at the time of writing, the second-highest-emitting country in the world based on shares of emissions, never ratified the Kyoto Protocol. It has therefore not been held to account for lowering emissions since the implementation of the Protocol in 2008 and will not be held to account until 2020 (if it remains part of the Paris Agreement). In addition, Canada, Japan and Russia (each in the top ten for global shares of emissions) have refused to participate in the second commitment period of the Kyoto Protocol. This means they are not held to account for lowering emissions between 2012 and 2020. Under a PATP model, all of these countries must be held to account because of their high levels of emissions.

When considering wealth levels, a similar story emerges. According to the World Bank (2016), the USA, who never ratified the Kyoto Protocol, had the highest Gross Domestic Product (GDP) in 2016, making it the richest country in the world. Japan, Canada, Russia and New Zealand, who have all refused to participate in the second round of the Kyoto Protocol, rank 3rd, 10th, 12th and 51st in terms of GDP. These are some of the richest countries in the world, and the fact that they are not being held to account for financial contributions under the Kyoto Protocol until 2020 is not in line with the PATP model of responsibility. Furthermore, China ranks 2nd and India ranks 7th in terms of GDP, strongly indicating that the category of 'less developed' under the UNFCCC definition seems to leave out some very rich countries from the scope of responsibility, which is not promising in terms of enabling the PATP. Overall, then, it is clear that the Kyoto Protocol does not hold many of the highest emitters and wealthiest countries to account. This is not compatible with the PATP model, which calls for countries to be held to account in line with their level of emissions and/or level of wealth. For these reasons, the actors under the UNFCCC cannot be said to have put policies in place that are consistently leading towards a context where Demand Two can be met. Instead, actors under the UNFCCC fall onto rung three of the four-point hierarchy, because they have made normative commitments (in the Convention and Kyoto Protocol) but not adopted policy that lives up to these:

> 3. Actors in the institution have promised to begin working on enabling the demand of justice in the future – no policy has been adopted, but there is the potential for the creation of policy in order to consistently work towards enabling the demand of justice.

The potential for the creation of policy to consistently work toward enabling Demand Two lies not only in the flexibility of the CBDR, but also in the fact that the Kyoto Protocol encompasses policies which, although not in line with the PATP, nevertheless indicate that there is an existing policy framework that can be built upon. However, the limited number of wealthy/high-emitting states currently held to account is a key hindrance that must be overcome in order to bring about a more just response to climate change. Encouragingly, negotiations concerning which countries are to be held to account for emissions reductions and/or financial contributions have been ongoing since 2007. These negotiations have been fraught with difficulties, and ultimately, although there was some hope that the Paris Agreement would change the categorisations of Annex I and II, this was not possible. However, there has been some important progress towards expanding the list of countries that are currently held to account for emissions reductions and financial contributions. To demonstrate the progress that has been made, the chapter will now briefly outline the negotiations that took place in the lead-up to the Paris Agreement before assessing the Agreement itself.

A key part of understanding the negotiations in the run-up to the Paris Agreement is to examine the role of the 'Ad Hoc Working Group on the Durban Platform for Enhanced Action' (referred to as the ADP by the UNFCCC), which was set up in 2012 and tasked with ensuring that negotiators had a text to work with at COP21. At COP19 in Warsaw, the ADP Parties discussed the process for defining mitigation commitments (ENB 2013: 10). There was disagreement over which countries should be responsible for how much, with less developed countries keen on basing mitigation targets on historical responsibilities, and developed countries arguing that historical responsibilities will not ensure achievement of the 2°C goal (ENB 2013: 11). This reflects old divisions that have been in place since the Convention was signed over twenty years ago, with less developed countries pushing for a PPP approach, and developed countries backing an ATP approach. During negotiations, less developed countries emphasised the continued application of the annex-based differentiation arrangement under the Convention and developed countries emphasised the need to continue but also update the application of the CBDR principle to reflect evolving circumstances (ENB 2013: 13). In the end, although the Parties agreed to encourage all member states to initiate or intensify domestic preparations for their Intended Nationally Determined Contributions, or INDCs, ahead of COP21 in Paris, key questions on how to differentiate commitments of developed and less developed countries were left unresolved (ENB 2013: 29).

At COP20 in Lima, ADP negotiations did not make much headway, because negotiators could not come to an agreement on how to allocate responsibilities for mitigation. Japan, New Zealand, the USA, Australia, Switzerland and Canada opposed creating binary divisions on commitments, based on annexes or the distinction between developed and developing countries (ENB 2014: 37). India with China, Brazil, Fiji, the Dominican Republic, Thailand and Venezuela, along with the 'Like Minded Group of Developing Countries' (LMDCs), all wanted to ensure that the new global agreement would not only be 'guided by', but instead be 'in accordance with', the principle of CBDR and provisions under the Convention, which indicated their continued commitment to existing categories of responsibility (ENB 2014: 32). China and Brazil, in particular, opposed the introduction of new categories, claiming that diverting from the principles and provisions of the Convention would make progress on a new global agreement difficult (ENB 2014: 37). The final draft agreement that came out of COP20, the 'Lima Call for Action', did not fully address these differences in opinion and left much more work to be done to find an agreement at COP21.

The ADP met three times in 2015. Their focus was on taking the 'Lima Action Plan' forward, and negotiators worked through the elements of the text section by section, proposing additions in places where they felt their views were not adequately reflected

(ENB 2015a: 1). However, although a text was eventually agreed to, the text itself did not contain any concrete decisions, instead outlining the positions of all Parties. This is largely because disagreements over how to classify countries and how much each country should contribute were not resolved (ENB 2015a: 4). Interestingly, the USA proposed a placeholder for a new 'Annex X', to be agreed at COP21 and updated regularly based on criteria relating to evolving emissions and economic trends. This is indicative of a PATP model that takes into account both wealth and emissions levels. By contrast, Iran, for the LMDCs, proposed noting that the largest share of current global GHG emissions originates from developed countries and that emissions in developing countries will grow to meet their social and development needs. This is more indicative of a PPP model, which is what less developed countries have advocated for decades. The differences in interpretations of CBDR were unfortunately not resolved by the ADP. Instead of agreeing on a proposal, each proposal was simply listed as an option in the agreed draft text that would form the basis of the Paris Agreement. According to analysts present at the 2015 ADP meetings, this was a necessary precondition to generate a sense of ownership among parties and boost confidence that all parties' views would be taken into consideration in the negotiations on the Paris Agreement (ENB 2015a: 13).

During the negotiations at COP21 Paris, the draft text was shortened significantly, and the options to change the categorisations which had been suggested by the ADP were ultimately removed. As a result, the Paris Agreement makes no explicit mention of the annexes or the historic indications of differentiation, but instead discusses developed and less developed countries as broad categories (ENB 2015c: 43). For example, Article 4.4 of the Paris Agreement states that 'developed country Parties should continue taking the lead by undertaking economy wide absolute emission reduction targets. Developing country Parties should continue enhancing their mitigation efforts and are encouraged to move over time towards economy wide emission reduction or limitation targets in the light of different national circumstances' (UNFCCC 2015: 22). In this sense, less developed countries are not held to account for immediate action, but rather encouraged to act over an unspecified period of time. In addition, developed countries are not bound by the Agreement to increase their mitigation or support efforts beyond existing commitments (ENB 2015c: 43). So, all in all, the Paris Agreement did not result in a significant shift or rethink of the Annex countries or the definition of the CBDR.

However, although no changes have been made to how countries are categorised, there has been a very significant shift in how responsibility is determined. While the ADP was discussing how to renegotiate responsibilities from 2012–15, countries were busy drafting their INDCs in the run-up to the negotiation of the Paris Agreement at COP21. At COP19 in Warsaw negotiators decided that all Parties should 'initiate or intensify' their efforts to submit INDCs. At COP20 in Lima, all Parties were invited to communicate their INDCs to the secretariat well in advance of COP 21. Over 160 countries submitted their INDCs in the run-up to COP21, and the total number of countries who have submitted to date (COP23) is 190. The INDC model represents a seismic shift in determining who is responsible for climate change action.

The post-Paris Agreement regime is now less about reaching a 'global deal' and more about a 'pledge and review' system (Hale 2016: 12). In a sense, the regime is now about self-differentiation rather than defining differences between countries in a top-down manner. As former USA Secretary of State John Kerry explained, the INDCs are a 'monument to differentiation': each country determines its 'fair contribution', according to its respective capabilities and in light of its 'different national circumstances'. This sounds very much like a nationally determined version of CBDR. Instead of dictating who is responsible for what action on the global level – something countries have been unable to agree on since the original Convention in 1992 – states are now allowed to make their own case for how much responsibility they will take on and why. As was discussed above, the INDCs will be reviewed in the global stocktake every five years, meaning that the concept of CBDR is likely to shift over time, as countries revise their commitments.

This new approach has, for the first time since the UNFCCC was founded, brought very large emitters into the mitigation regime (Hale 2016: 12). There are currently 190 countries who have submitted INDCs, representing 95% of global emissions (ENB 2015c: 44). Although these INDCs are so far inadequate in terms of keeping global emissions at a level which does not threaten the right to health, as discussed above, the fact that 190 countries have submitted plans is promising in terms of enabling a PATP model of responsibility. The list includes China, Brazil, Canada and New Zealand, who are not currently held to account under the Kyoto Protocol, and until very recently also included the USA.

At the time of writing, it is not yet clear whether the USA will partici-pate in the Paris Agreement. Nevertheless, all in all, the INDCs extend the list of countries that are responsible for lowering emissions from under forty (under the Kyoto Protocol) to nearly every country in the world, which is a significant increase. Of course, it remains to be seen how much pressure is put on high-emitting less developed countries such as China and Brazil once the Paris Agreement is in force, and what enforcement measures will be put in place to ensure compliance. At the moment, although the communication of INDCs is legally binding, their content and targets are not (ENB 2015c: 43).

The shift to INDCs also indicates a shift in how less developed countries have been included in decision-making. Although their pre-ferred PPP interpretation of CBDR was ultimately not included in the Paris Agreement, these countries have more of a say in the multilateral regime than ever before. This can be seen in their role in the ADP nego-tiations, where their ideas were included alongside those of developed countries right up until the Paris Agreement, when all the suggestions for change were ultimately removed. In addition, the introduction of INDCs, as well as NAPAs/NAPs, and a dedicated Adaptation Frame-work (discussed above) all seem indicative of a transforming regime.

Less developed countries are increasingly reporting their needs, achievements and plans alongside developed countries. Hopefully this will ultimately mean they play a stronger role in decision-making pro-cedures. Of course, there is still a need for a coordinated approach to increase the negotiation skills and technical capacity of less developed country negotiators (Abeysinghe and Huq 2016: 199). There is hope for this happening in future, especially considering that the Least Developed Country Work Programme aims to 'provide training, on an on-going basis, in negotiating skills and language, where needed, to develop the capacity of negotiators from Least Developed Countries to participate effectively in the climate change process' (Abeysinghe and Huq 2016: 199). This will be further discussed in the final section of this chapter. For now, the chapter turns to the third and final demand of climate justice.

DEMAND THREE – THE RESPONSIBILITIES OF NON-STATE ACTORS

Demand Three states that all capable actors, including individuals, firms, sub-state entities and international institutions, irrespective of the country in which they live or exist, must be held responsible for lowering emissions and/or contributing to adaptation efforts, in line with their respective capabilities. In this sense, to enable Demand Three, actors in the UNFCCC must create a context where non-state actors are held to account for their climate-related responsibilities. The role of multilateral governance in this matter is highly important, because it is by now very much clear that individual actors are unlikely to constrain their consumption patterns without actions by governments that push them to do so (Harris and Lee 2017: 5). In addition, although non-state actors have taken on a substantive role in addressing the climate change problem, as will be discussed in the next chapter, these actors are mostly unregulated and acting of their own accord on a voluntary basis. This is potentially problematic because action on many scales and through many dispersed but overlapping non-state actors has the potential to be ineffective (Bulkeley and Newell 2010: 106). It may therefore be necessary to regulate these actors in order to enable them to act on their responsibilities most effectively. This will be further discussed in the conclusion of the book, but for now, the current chapter aims to investigate to what extent multilateral actors have enabled non-state actors to meet their climate-related responsibilities.

Let us begin by considering the Convention. The preamble 'recalls that states have [. . .] the responsibility to ensure that activities within their jurisdiction or control do not cause damage to the environment of other states or of areas beyond the limits of national jurisdiction' (UNFCCC 1992: 2). The wording in this passage places responsibility for climate change action exclusively on states. Holding states to account is not, in its own right, problematic for the fulfilment of Demand Three. Capable actors, including individuals, firms, sub-state entities and international institutions, can be held to account within states, through regulatory measures such as taxation or environmental law. However, as was discussed above, the preamble of the Convention places the burden of responsibility solely on a limited number of developed states in which 'the largest share of historical and current global emissions of greenhouse gases has originated' (UNFCCC 1992: 2). This is problematic, because any capable actor outside of these states, including individuals, firms, sub-state entities and international institutions, escapes responsibility and regulation here.

The legally binding body of the Convention is in line with the tone set in the preamble. Article 3.1 of the Convention states that 'the Parties should protect the climate system . . . . in accordance with their common but differentiated responsibilities and respective capabilities.

Accordingly, the developed country Parties should take the lead in combating climate change and the adverse effects thereof' (UNFCCC 1992: 9). This is problematic, because non-state actors outside of these countries escape responsibility and regulation. As was discussed in Chapter 3, there are rich individuals with high emissions living in countries such as India and China, states that are not held to account under the Convention. Furthermore, there exist international institutions and corporations outside of developed countries which are capable of financial contributions; for example, Petro China, or Telemar in Brazil, who earn billions in revenue every year.

Even more problematically, the Convention makes no mention of non-state actors, full stop. This is perhaps unsurprising, given that these actors only emerged on the scene over the past decade and were feasibly not seen as important in 1992, when the Convention was signed. Nevertheless, the Convention forms the basis of climate change action from 1992 onwards, and the omission of non-state actors is therefore potentially quite problematic. It sets a tone that has proved difficult to shake until very recently, as will be explained below. In sum, then, the Convention cannot be said to represent ambitions in line with Demand Three, which implies that the actors under the UNFCCC do not enable Demand Three of justice, because these actors do not create a context where Demand Three can be met. In other words, according to the above assessment of the Convention, the actors in the UNFCCC fall under the fourth rung of the four-point hierarchy:

4. Actors in the institution do not enable the demand of justice – there has been no promise or attempt to enable the demand of justice and there are no policies in place.

In order to investigate whether the actors under the UNFCCC have made a radical departure from the Convention as non-state actors gained importance on the global stage, the Kyoto Protocol and Paris Agreement will now be examined. Interestingly, although the Kyoto Protocol does not present a significant change from the original Convention, the negotiations in the run-up to the Paris Agreement and the Paris Agreement itself represent a substantial shift in policy in terms of including non-state actors in multilateral climate change action. Let us begin by examining the Kyoto Protocol, which reflects the Convention almost exactly. Although Article 10 of the Kyoto Protocol stipulates that 'all Parties, taking into account their common but differentiated responsibilities ... shall formulate programs to improve the quality of local emission factors and report on these' (UNFCCC 1998: 9), the Kyoto Protocol, in line with the Convention, exclusively holds developed countries to account for lowering emissions and financial contributions. As was explained above, countries in Annex I are held to account for emissions reductions, and those in Annex II for contributing to climate change costs.

This allocation of responsibilities is not in line with Demand Three, which states that *all* capable agents must be held to account. Under the Kyoto Protocol, which is in force until 2020, just under forty states are held to account for climate change action. Furthermore, as was mentioned above, rich and high-emitting states such as the USA never ratified the Kyoto Protocol, and Canada, Japan, Russia and New Zealand have refused to participate in the second commitment round (2013–20). This lowers the number of states held to account for financial contributions and emissions reductions under the Kyoto Protocol further still. Although the states that are held to account under the Kyoto Protocol can hold individuals, firms, sub-state entities and international institutions which exist within their borders to account, this still leaves well over one hundred states which are not held to account. Many of these states have capable (rich and high-emitting) individuals, firms (Shell, Exxon for example), and sub-state entities (cities like Mumbai, Beijing, New York, Toronto) present within their borders. For this reason, the Kyoto Protocol cannot be said to implement policies that hold all capable actors to account. In this sense, multilateral actors do not enable Demand Three through the Kyoto Protocol, because they do not create a context where Demand Three can be met, and therefore continue to reside on the bottom rung of the four-point hierarchy, at least in terms of the Kyoto Protocol:

> 4. Actors in the institution do not enable the demand of justice – there has been no promise or attempt to enable the demand of justice and there are no policies in place.

Although this is no doubt discouraging, there have recently been two very important shifts in multilateral policy which suggests that Demand Three may be enabled in the near future. First, as was discussed above, 190 countries have now submitted Intended Nationally Determined Contributions (INDCs), which means more countries than ever have pledged to take action on climate change. This, in turn, could mean that non-state actors, including individuals, firms, sub-state entities and international institutions will be held to account within these countries. Indeed, many of the INDCs include discussions of how such actors will be part of climate change action (Sander et al. 2015: 469). Instead of seeing non-state actors as alternatives to or substitutes for national and intergovernmental commitments, they are increasingly seen as both complements to and 'means of implementation' for INDCs (Sander et al. 2015: 469). India's INDC, for example, has a section dedicated to a discussion of how individuals and the private sector can and will contribute to their climate change efforts. It seems that the bottom-up INDC regime is allowing multilateral actors to include more non-state actors in the scope of climate change action. This is encouraging, because it creates a context where capable actors can be held to account for lowering emissions and/or making financial contributions, in line with Demand Three.

The second important shift comes in the form of changing policy towards non-state actor inclusion at the top-down level of the multilateral climate change response. This change in policy has come about very recently in the run-up to the negotiation of the Paris Agreement, under the leadership of the 2014 Peruvian and 2015 French presidency of the UNFCCC, respectively. The change in policy is perhaps unsurprising, given that the growth of non-state climate change action is increasingly impossible to ignore, and that many groups have been actively lobbying governments for greater inclusion and recognition in the UNFCCC process for the past decade (Hale 2016: 14). Nevertheless, it is worth briefly reflecting on how change has come about so quickly. The first sign of changing times could be seen in September 2014, when former United Nationals Secretary General (UNSG) General Ban Ki-Moon organised a 'Climate Summit'. What was different about this event to previous ones is that he invited not only heads of state, but also CEOs, mayors and other non-state leaders, to take on concrete commitments regarding climate mitigation, adaptation and finance (Hale 2016: 14). Although this was not an official UNFCCC event, it signalled that there was momentum for change within the multilateral regime, because the UNSG was keen to include non-state actor voices in discussions on climate change.

In December 2014, just a few months after the 'Climate Action' summit, COP20 in Lima took place. As has become customary in recent years, there were non-state actors taking part in side events separate to the multilateral negations. However, there was a crucial difference this year, because non-state actors were invited to a 'High Level Action Day' as part of the negotiations to make announcements and commitments, channelling the spirit of the New York Summit (Hale 2016: 15). In addition, non-state actors were discussed as part of the draft agreement for COP21 in Paris for the first time. The Peruvian presidency, working with the UNFCCC Secretariat, included numerous references to cities and the private sector in the draft text to be adopted in Lima. Unfortunately, opposition and inertia in the negotiating process meant that all references to non-state actors were stripped from the text on the last night of negotiations (Hale 2016: 15). Opposition came mainly from the G77, with some countries like Sudan worried that references to non-state actors could undermine state sovereignty, and other countries, like Venezuela, concerned about the prominence that would be given to multinational corporations (Hale 2016: 15).

Despite this setback, Christiana Figueres, the president of the UNFCCC at the time, and Ban Ki-Moon, the UNSG at the time, were eager to continue the strategy of mobilising non-state actors and including these actors in multilateral negotiations. Together they formed the 'Lima-Paris Action Agenda' or LPAA in late 2014, with the aim of bringing non-state actors 'inside the COP walls' (ENB 2015b: 45). The LPAA sought to orchestrate and expand climate initiatives from non-state actors, and eventually came to include over seventy initiatives containing over 10,000 commitments or actions by cities, companies, states and others, encompassing twelve thematic areas, such as cities, forests, energy efficiency and renewables, with a lead partner responsible for mobilisation in each (Hale 2016: 15). In order to aid mobilisation of action, the 'Nonstate Actor Zone for Climate Action' or NAZCA portal was created to track climate actions and pledges from all sectors of society. The NAZCA aggregates data on climate actions at all levels, resulting in improved visibility of non-state and sub-national initiatives (Sander et al. 2015: 468). In addition, although all references to non-state action had been cut out at Lima, they were brought back in subsequent versions of the COP21 negotiating text, with the French and Peruvian governments, along with the UNFCCC Secretariat, ensuring that successive drafts throughout 2015 included language recognising and encouraging cities, companies and other non-state actors' efforts.

At COP21 in Paris, the French presidency declared non-state action the 'fourth pillar' of the COP21 negotiations, alongside, and equal to, the national pledges, the financing package and the negotiated agreement (Hale 2016: 14). Negotiations at COP21, for the first time, included the announcement of transnational initiatives on themes like forests, cities or energy efficiency (Hale 2016: 14). The efforts of the LPAA in the run-up to Paris, and the inclusion of non-state actors in negotiations, had a significant effect on the outcome of the Paris Agreement. The negotiation text explains that Parties 'agree to uphold and promote regional and international cooperation in order to mobilise stronger and more ambitious climate action by all Parties and non-Party stakeholders, including civil society, the private sector, financial institutions, cities and other subnational authorities, local communities and indigenous peoples' (UNFCCC 2015: 2). Furthermore, the negotiation text encourages Parties 'to work closely with non-Party stakeholders to catalyse efforts to strengthen mitigation and adaptation action' (UNFCCC 2015: 17). The intention to include these actors in multilateral climate change action, and to encourage states to pursue collaborations with these actors, is very clear here. More than merely welcoming action, non-state actors are invited to 'scale up their efforts and support actions to reduce emissions and/or to build resilience and decrease vulnerability to the adverse effects of climate change and demonstrate these efforts via the Non-State Actor Zone for Climate Action platform' (UNFCCC 2015: 19). Overall, then, it is very clear that the intention of the post-Paris Agreement regime is to view non-state actors not as an alternative to the UNFCCC process, or as merely a helpful addition, but as a core element of its logic of spurring rising action on climate over time (Hale 2016: 14).

Interestingly, the legally binding Articles of the Paris Agreement make no mention of non-state actors. The only part where they are explicitly referred to is in the preamble, where 'the importance of the engagements of all levels of government and various actors' is reiterated (UNFCCC 2015: 21). The wording here is quite vague, suggesting that perhaps not every country shares the ambition to include non-state actors in the post-Paris Agreement regime. This leaves open the question of what is next, or more exactly the question of whether multilateral actors will enable Demand Three by creating a context where non-state actors can be held to account for their climate-related responsibilities. Governments are currently considering how non-state initiatives will develop and be institutionally supported in the post-Paris Agreement period as part of their INDCs (Sander et al. 2015: 468). Overall, though, it seems that COP21 made substantial – though still incomplete – progress toward an effective framework for non-state climate action. The negotiations in Paris opened an unprecedented window of opportunity in which to build a constructive new relationship between multilateral and non-state initiatives (Sanders at al. 2015: 469). Taking up this opportunity will require governments to sustain political support beyond COP21, as well as adequate resources to ensure that the framework for non-state actors can deliver on its potential (Sander et al. 2015: 469). This will be further discussed in the concluding section of the chapter.

CONCLUSION: SUMMARY OF FINDINGS

The chapter has now completed its assessment of the multilateral climate change response. What is perhaps most striking about this assessment is that none of the three demands of climate justice have been fully enabled. In fact, actors sit on the third rung of the four-point hierarchy in the case of Demands One and Two, and on the fourth rung in the case of Demand Three:

The Four-Point Hierarchy

1. Actors in the institution enable the demand of justice – the demand of justice is unequivocally fulfilled in its entirety.
2. Actors in the institution are consistently working towards enabling the demand of justice – the demand of justice is not yet fulfilled, but there are policies in place which are consistently leading towards this goal.
3. Actors in the institution have promised to begin working on enabling the demand of justice in the future – no policy has been adopted, but there is the potential for the creation of policy in order to consistently work towards enabling the demand of justice.
4. Actors in the institution do not enable the demand of justice – there has been no promise or attempt to enable the demand of justice and there are no policies in place.

Although these results are not indicative of a just multilateral climate change regime, it is nevertheless encouraging to see that multilateral actors have moved beyond the bottom rung of the hierarchy in the case of Demand One and Two and have made plans that are in line with Demand Three in the run-up to the Paris Agreement. It is worth pausing briefly to appreciate the fact that multilateral actors' promises and ambitions are very much in line with the three demands of justice set out in this book. As Harris (2010: 75) has optimistically put it, the multilateral climate change response 'may be at or near the zenith of justice in international environmental affairs, demonstrating that most states recognise the need for justice in this issue area'. While this is quite a rosy view of climate change governance, the assessment conducted in this chapter confirms that the rhetoric of justice is inscribed in the framework of the UNFCCC. The Convention, in particular, appears to reveal that the actors under the UNFCCC view climate change as a matter of global justice. And, as was demonstrated above, multilateral actors have gone beyond making promises and have attempted to implement policies which fulfil the ambitions set out in the Convention in both the Kyoto Protocol and the Paris Agreement. These policies are, of course, so far inadequate, but it is clear that multilateral actors are consistently moving forward. This is most clearly demonstrated by the Paris Agreement, which demonstrates movement forward on each demand of justice:

The Three Demands of Climate Justice

1. a) Global temperature changes must be kept at or below 2°C.
   b) Adaptation must be prioritised alongside mitigation.
2. The distribution of benefits and burdens in global climate change action must be based on the Polluter's Ability to Pay (PATP) model.
3. Capable actors, including individuals, firms, sub-state entities and international institutions, irrespective of the country in which they live or exist, must be held responsible for lowering emissions and/ or contributing to adaptation efforts, in line with their respective capabilities.

In terms of the first part of Demand One, the Paris Agreement has resulted in more countries than ever before pledging to lower emissions, which means that the right to health has a higher chance of being protected than under the Kyoto Protocol. In terms of the second part of Demand One, the multilateral regime has made progress in terms of monitoring adaptation needs through NAPAs/NAPs and INDCs, and the Paris Agreement appears to move forward on adaptation policy, at least in terms of what has been promised. The creation of the Least Developed Experts Group (LEG) and Adaptation Committee are also encouraging, because they create space for adaptation to be prioritised alongside mitigation.

In terms of Demand Two, although the Paris Agreement was not able to move forward on the outdated categorisation of Annex I and II, the INDCs have, for the first time, ensured near-universal participation in the climate change regime. There are now more rich and high-emitting countries being held to account than ever before, albeit under weak enforcement mechanisms. The focus on INDCs and NAPAs/NAPs rather than a top-down regime also indicates a shift in how less developed countries have been included in decision-making. Although their preferred interpretation of CBDR was ultimately not included in the Paris Agreement, these countries have more of a say in the multilateral regime than ever before.

Finally, in terms of Demand Three, the run-up to the Paris Agreement and the negotiations at COP21 represent a significant shift in the perception of the role of non-state climate change actors. Although the Paris Agreement itself does not make mention of these actors, it was clear from the negotiations that non-state actors will play an increasingly important role. Many of the INDCs make mention of non-state actors, and it is clear from the assessment above that multilateral actors are beginning to shift their attitude towards these actors. Of course, taking up the opportunity to include these actors will require states to sustain the political support created in Paris beyond COP21. Overall, though, there has been significant movement forward on all three demands of justice.

Nevertheless, the fact that actors under the UNFCCC are failing to fully enable any of the three demands represents a key hindrance to a just climate change response. The actors under the UNFCCC must do more to live up to their responsibility for enabling a condition of justice. This is an opinion that is shared by cosmopolitan scholars who have previously assessed the multilateral regime. Lawrence (2014: 126), for example, argues that the relatively robust, albeit vague, justice principles reflected in the Convention contrast with the lack of effective and binding emission targets and other essential elements such as compliance mechanisms and funding necessary for an effective global regime. Similarly, Gardiner (2004b: 36) points out that the Kyoto Protocol is no more than a comfortable illusion that serious progress is being made. Harris (2010: 75) agrees, noting that while the multilateral regime may be considered an important step for international justice, in practice multilateral governance has not achieved the aims of preventing dangerous impacts on the earth's climate system. These assessments are rather pessimistic and do not imply much hope for the future. Unfortunately, they seem quite accurate.

In terms of Demand One, the targets set out in the Kyoto Protocol are indisputably not in line with what is required to protect the human right to health. The small number of countries held to account until 2020 only account for 15% of global emissions, and their inadequate efforts will not make enough impact to protect future generations. In addition, although the Paris Agreement has increased ambitions by setting a 1.5°C temperature change goal, the current INDCs are not in line with this target. As was explained above, global temperatures are set to warm by between 2.7°C and 4.9°C, reaching 3.6°C by 2100 under the current INDCs. Furthermore, although the Paris Agreement has set out new compliance mechanisms, these remain voluntary and non-punitive, calling into question to what extent states will be held to account for their INDCs. The story is largely the same for adaptation. Although the creation of an Adaptation Committee, LEG and the focus on NAPAs are all signs of progress, the funding mechanisms set out in the Kyoto Protocol have so far failed to raise adequate finance for adaptation, and less developed countries remain frustrated at the slow pace of technology transfer. In addition, NAPAs have not yet been acted on, and the Paris Agreement did not set out new financial targets. Furthermore, financial contributions and technology transfer remain voluntary, calling into question whether the pace of adaptation will ever be adequate for the protection of the human right to health.

In terms of Demand Two, the Kyoto Protocol unequivocally does not enable the PATP model of responsibility, because it fails to include many of the richest and highest-emitting countries. And, although the Paris Agreement has moved forward on this problem with the implementation of INDCs, there are still disagreements over who should be responsible for what, with developed countries favouring an ATP model and less developed countries preferring a PPP model. It is unclear which model will be pursued, and to what extent rich and high-emitting countries will be expected to contribute more than poorer and low-emitting countries. Furthermore, although less developed countries are increasingly able to report on their needs, their voices are still often left out of climate change negotiations. There is hope for improvement on this in future, especially considering that the Least Developed Country Work Programme aims to 'provide training, on an on-going basis, in negotiating skills and language, where needed, to develop the capacity of negotiators from Least Developed Countries to participate effectively in the climate change process' (Abeysinghe and Huq 2016: 199). However, to what extent they will be included in negotiations in the run-up to the Paris Agreement being implemented remains to be seen.

Finally, in terms of Demand Three, neither the Convention nor the Kyoto protocol adequately hold non-state actors to account, or even make mention of these actors. And, although the negotiations at COP21 included non-state actors more seriously than ever before, the Paris Agreement ultimately made no mention of these actors. This leaves open the question of what is next, or more exactly the question of whether multilateral actors will enable Demand Three by creating a context where non-state actors can be held to account for their climate-related responsibilities. Multilateral actors are currently considering how non-state initiatives will develop and be institutionally supported in the post-Paris Agreement period as part of their INDCs (Sander et al. 2015: 468). Taking up this opportunity will require governments to sustain political support beyond COP21, as well as adequate resources to ensure that the framework for non-state actors can deliver on its potential (Sander et al. 2015: 469). Whether this will happen remains to be seen in the run-up to the Paris Agreement being implemented. Overall, then, multilateral actors have certainly made progress, but also have a long way to go to ensure a just response to the climate change problem.

Even though multilateral actors have not yet enabled a just climate change regime, Harris, Gardiner and Lawrence all made their assessments before the Paris Agreement was ratified. As was discussed above, the Paris Agreement has resulted in movement forward on each demand of justice. In this sense, the mention of 'climate justice' in the preamble of the Paris Agreement is not simply empty rhetoric (UNFCCC 2015: 21). Of course, the Paris Agreement is not perfect, and policies are still not in line with what is required for a just climate change response. Nevertheless, the outcome of COP21 exceeded expectations, producing an agreement that while perhaps not a revolution, is an important step in the evolution of climate governance and a reaffirmation of environmental multilateralism (ENB 2015c: 43). In the end, the success of the Paris Agreement will hinge on its ability to encourage multilateral actors to ratchet up their contributions to a sufficient level of ambition to safeguard the planet (ENB 2015c: 44). Post-2020 ambition relies heavily on pre-2020 ambition. As multilateral actors put it during the closing plenary at COP21, 'the work starts tomorrow' (ENB 2015c: 44). Paris truly 'was more of a beginning than an end' (Hale 2016: 20).

In order to gauge how this work is going, the Conclusion of this book will briefly discuss the latest state of climate change negotiations, commenting on the extent to which there has been movement forward at COP22 and COP23. Before this can be discussed, the book must assess to what extent actors involved in transnational climate change governance enable a condition of justice in Chapter 6. This is necessary due to the conception of responsibility outlined in Chapter 4, which argued that the actors under the UNFCCC have formal authority to act and are therefore more responsible for enabling a condition of justice in the case of climate change. However, it was explained that this does not diminish the moral responsibility of other actors, specifically those involved in transnational governance processes, if the actors under the UNFCCC should fail to enable the three demands of justice explicated in this book. As this chapter has clearly illustrated, actors under the UNFCCC do not fully enable any of the demands of justice. Therefore, the book now moves onto the assessment of actors involved in transnational climate change governance.

NOTES

1. Compared to an average rate of 3.5% in the 2000s and 1.8% over the most recent decade, 2006–15.
2. For a Discussion of the PATP, see Chapter 3.
3. For a discussion of the PPP, ATP and PATP, see Chapter 3.

Chapter 6

ASSESSING TRANSNATIONAL
CLIMATE GOVERNANCE

INTRODUCTION

The second half of this book focuses on the cosmopolitan assessment of the global response to climate change, both multilateral and transnational. Now that the book has assessed the multilateral climate change response, it turns to assessing transnational responses. Chapter 4 explained that both multilateral actors under the United Nations Framework Convention on Climate Change (UNFCCC) and actors involved in transnational climate change governance processes have a moral responsibility to enable a condition of climate justice, due to their capability of restructuring the social and political context so that the three demands defended in this book can be met. Furthermore, it was explained that although multilateral actors are most responsible for enabling a condition of justice in the case of climate change, transnational actors must act on their responsibilities if multilateral actors fail to enable a condition of climate change justice. The previous chapter assessed the multilateral climate change response and argued that it has unequivocally failed to fully enable any of the three demands of justice set out in this book. This failure implies that transnational climate change governance actors have a responsibility to enable these demands and thereby facilitate a just response to the climate change problem.

The purpose of the current chapter is therefore to assess to what extent transnational actors enable a condition of justice in the case of climate change. Further to this, the chapter aims to provide a broad-brush overview of transnational climate change governance. This will allow readers to develop an understanding of the vast and rapidly changing non-state climate change response. To provide this overview, the chapter makes use of both existing climate change governance research and

161

ten examples of transnational climate change governance initiatives. This allows the chapter to explain how climate change governance has developed and where it stands today. Mirroring the previous chapter, the current chapter will focus on one demand of climate justice at a time, assessing both what has been promised by transnational actors and what has been achieved so far. The final part of the chapter summarises the findings made, compares them to those made in the previous chapter and considers what role multilateral and transnational actors might play in the post-Paris Agreement climate change regime. This will be expanded on in the conclusion of the book.

TRANSNATIONAL CLIMATE GOVERNANCE: A JUSTICE-BASED EVALUATION

Before the assessment of transnational climate change governance can commence, the chapter must briefly explain how this assessment will be conducted. Although some scholars have been optimistic about transnational climate change governance, claiming that transnational actors have considerable potential to mitigate climate change, help affected communities adapt to its effects and leverage financial and other resources (Sander et al. 2015: 467), these claims are difficult to verify. Assessing transnational climate change governance actors is not as straightforward a task as assessing actors under the UNFCCC, who operate under a common legal framework and aim to create global agreements around climate change. Transnational climate change governance actors are not focused on a single outcome or operating under an overarching framework, but rather pursue a variety of climate change responses that are very much independent from one another. This makes it difficult to compare initiatives or to make definitive claims about the transnational climate change response. In addition, the number of transnational initiatives grows exponentially each year (Abbott 2012: 571). It is therefore challenging to discern the 'total universe' of transnational climate change governance activities (Bulkeley et al. 2015: 19). In fact, the exact number of transnational climate change governance initiatives is unknown (Hoffman 2011: 25), although it is estimated that the numbers of projects in existence is in the thousands (Stevenson and Dryzek 2014: 87). So, although some scholars, such as Bulkeley et al. (2015: 9), aim to provide a broad overview of transnational climate change governance, they explain that it would be impossible to find a 'representative' or even 'random' sample of non-state climate change responses. To make matters even more complicated, most transnational climate change governance initiatives are relatively new and are experimental, which means we cannot yet be sure how they will turn out (Hoffman 2011: x).

It is perhaps for this reason that there has been limited systematic analysis of transnational climate change governance (Bulkeley et al. 2015: 11). Examples of this type of analysis include Hoffman's (2011) monograph, which provided an assessment of fifty-eight climate governance initiatives, Abbott's (2012) research paper, which mapped sixty-seven transnational organisations and Bulkeley et al.'s (2015) evaluation of sixty examples. Bulkeley et al.'s project provides the most comprehensive overview of transnational climate change governance to date. These scholars have, for the first time, assessed transnational climate change governance from three perspectives: agency, special and system dynamics, and critical political theory. Although this assessment is important, there has so far been no attempt to comprehensively assess transnational climate change governance from one perspective.

Cosmopolitan assessment of transnational climate change governance is particularly lacking, with only a handful of scholars very recently attempting to begin such analysis in short papers or chapters (Bulkeley, Edwards and Fuller 2014; Hale 2016; McKendry 2016). The current chapter aims to build on this existing work by providing a global-justice-focused assessment of transnational climate change governance that can be compared to the assessment of multilateral actors in the previous chapter. As was explained in the Introduction of the book, such an assessment is necessary and important, because it has now become clear that transnational actors have the potential to contribute to the climate change response in a significant manner (Hsu et al. 2015). Their role is especially important at this present moment because the Paris Agreement presents a significant shift away from a purely state-led approach, with transnational climate change actors no longer being seen as a 'helpful addition' but as a core element of action on climate change (Hale 2016: 14).

To overcome the difficulty of assessing transnational climate change governance, the chapter will conduct a climate-justice-focused meta-analysis of the recent findings made by transnational climate change governance scholars. This existing research lies firmly outside of the climate justice field and has been conducted by a number of different scholars who have diverse areas of focus. The chapter aims to make sense of their disparate findings by assessing to what extent transnational actors have enabled a condition of climate justice. This analysis allows the chapter to paint a 'big picture' of what has happened so far, providing an overview of the development of transnational responses as well as explaining where they stand today. To aid the analysis, the chapter will explore ten examples of transnational climate change governance initiatives: the C40, the Climate Group, the Asian Cities Climate Change Resilience Network, 100 Resilient Cities, Carbon Sequestration Leadership Forum, the Asia Pacific Partnership on Clean Development and Climate, the Verified Carbon Standard, the Regional Greenhouse Gas Initiative, Carbon Trade Watch and Transition Towns.

These ten examples are not definitively representative of transnational climate change governance (because this is not possible) but have instead been selected due to their relevance to the assessment and their prominence within existing research of transnational climate change governance. This ensures that the assessment below ties in well to existing research on the subject. Furthermore, the ten examples include a wide range of actors, including states, cities, corporations, non-governmental organisations and individuals, as well as a wide range of governance mechanisms, including information sharing, monitoring, reporting, market mechanisms, capacity building and rule-setting. In this way, the examples, although not representative or all-encompassing, serve to provide an overview of a range of actors, mechanisms and types of initiatives involved in transnational climate change governance.

The chapter will now take each demand of justice in turn and assess to what extent transnational climate change responses enable this demand. The assessment will make use of the four-point hierarchy developed in Chapter 4. Both the demands and the hierarchy are outlined below, as a reminder to the reader:

The Three Demands of Climate Justice

1. a) Global temperature changes must be kept at or below 2°C.
   b) Adaptation must be prioritised alongside mitigation.
2. The distribution of benefits and burdens in global climate change action must be based on the Polluter's Ability to Pay (PATP) model.
3. Capable actors, including individuals, firms, sub-state entities and international institutions, irrespective of the country in which they live or exist, must be held responsible for lowering emissions and/ or contributing to adaptation efforts, in line with their respective capabilities.

The Four-Point Hierarchy

1. Actors in the institution enable the demand of justice – the demand of justice is unequivocally fulfilled in its entirety.
2. Actors in the institution are consistently working towards enabling the demand of justice – the demand of justice is not yet fulfilled, but there are policies in place which are consistently leading towards this goal.
3. Actors in the institution have promised to begin working on enabling the demand of justice in the future – no policy has been adopted, but there is the potential for the creation of policy in order to consistently work towards enabling the demand of justice.
4. Actors in the institution do not enable the demand of justice – there has been no promise or attempt to enable the demand of justice and there are no policies in place.

DEMAND ONE – PROTECTING THE HUMAN RIGHT TO HEALTH

Demand One focuses on the protection of the human right to health and states that: a) global temperature changes must be kept at or below 2°C and b) adaptation must be prioritised alongside mitigation. Although keeping global temperatures at or below 2°C seems increasingly unlikely due to current inaction, the IPCC maintains, at the time of writing, that there are multiple mitigation pathways which are likely to limit warming to below 2°C: emissions will have to be cut by 40%– 70% by 2050 compared to 2010, and will need to be near zero or below in 2100 (IPCC 2014a: 14). The chapter will therefore assess to what extent the transnational climate change actors are working towards these mitigation requirements by analysing existing research and exploring two key examples: the C40 and the Climate Group.

Assessing whether adaptation is being prioritised alongside mitigation is slightly less straightforward. It has been estimated that global adaptation costs will be $125 billion USD in 2050 (Hof et al. 2010: 252). So, one aspect to examine is the extent to which transnational actors have been working towards raising these funds. However, adaptation is not merely a matter of financial cost – it is also about sharing technology and ensuring that countries vulnerable to climate change are assisted in preparing for climate change. As was discussed in Chapter 3, these aspects of adaptation are particularly important for the right to health because strengthening health systems through technology and knowledge transfer, rather than financial transfer alone, could significantly reduce the burden of disease (WHO 2013: 1). The discussion below will bear the complexities of adaptation in mind when evaluating to what extent transnational actors are enabling the second part of Demand One. To aid the assessment, the chapter will analyse existing research as well as examining two key examples: the Asian Cities Climate Change Resilience Network and 100 Resilient Cities.

Mitigation

Measuring transnational climate change actors' emissions reductions is extremely difficult. In fact, some scholars claim that the long-term effects of transnational climate change governance processes on greenhouse gas (GHG) levels are impossible to estimate (Pattberg 2010a: 281). As Bulkeley et al. (2015: 159) put it, 'frustratingly, [GHG emissions] may be the worst metric to apply' to assess transnational climate change governance. This is mainly because many, if not most, initiatives are not directly aiming to reduce emissions. Instead, transnational actors are usually focused on providing information, incentives and capacity building for other actors to reduce their own emissions. Emissions reductions from these kinds of activities are difficult to measure. For example, transnational projects might focus on energy efficiency, changes in transportation and infrastructure or development and deployment of green technology. All of these efforts will have an indirect impact on emissions reductions, but this is not easy to measure because it is difficult to pinpoint which action influenced how much of a reduction in emissions. As another example, transnational projects may catalyse a change, perhaps by illustrating that a new technology is viable through a pilot project, leading to widespread adoption of this technology and resulting in significant emissions reductions. These reductions are difficult to attribute to the initial project. The problem lies in finding a direct connection between catalyst and amount of change. Such connections are nigh-on impossible to determine or indeed measure.

Furthermore, even when transnational actors explicitly aim to reduce emissions, they rarely set emissions targets. When assessing commitments made by 1,073 transnational actors at the 2015 New York Climate Summit, Hsu et al. (2015: 501) found that only eight included explicit emissions mitigation targets. And, in the rare cases that transnational actors do set out specific targets, it is difficult to measure whether these targets have been met. Researchers must often rely on self-reporting, because transnational actors are rarely monitored by a third party. For this reason, although there is some evidence of leadership on climate action, evidence of concrete, measurable impacts on emissions is scarce (Bulkeley et al. 2015: 161). Unfortunately, it is no easier to measure the cumulative impact of transnational climate change responses. Initiatives greatly differ in scale, scope and ambition, which makes it difficult to compare these projects across the board. Furthermore, a lack common baselines and the absence of strong monitoring mechanisms means that assessing a collective transnational impact on emissions is nearly impossible (Bulkeley et al. 2015: 160). Data on the level of emissions reductions resulting from transnational climate change governance projects is therefore almost non-existent.

All of these problems render it impossible to definitively claim that transnational climate change actors enable the first part of Demand One. If transnational actors are not setting emissions targets and their mitigation efforts cannot be measured or compared, then it is difficult to determine whether a context where emissions are lowered is being created. The lack of evidence on emission reductions has led some critics to claim that transnational climate change governance activity is merely a distraction from the hard work that is occurring at the multilateral level (Bulkeley et al. 2015: 158). In addition, some sceptics of transnational climate change governance claim that the transnational actors are simply about 'greenwashing': it is more important for these actors to show that something is being done, rather than make an actual impact on emissions (Bulkeley et al. 2015: 160). However, this dismissal of transnational climate change governance is too hasty. Considering that many initiatives are still relatively new, it may be too early to tell, definitively, whether this type of governance represents a case of greenwashing or whether actors within these processes are able to create a context that allows for significant emissions reductions.

Since concrete evidence is somewhat lacking, it is, for the time being, necessary to focus instead on the capacity of transnational climate change governance actors to create a context where emissions are reduced. Encouragingly, doing so leaves room for optimism. It should by now be obvious to readers of this book that an effective climate response will need broad participation. Transnational climate change governance actors have the capacity to create a context where this is possible, because these processes operate across multiple scales and engage a wide range of actors in the global response to climate change (Bulkeley et al. 2015: 158). This flexibility has allowed for an independent and creative pursuit of climate change responses in a manner that was simply not possible before. Unconstrained by the consensual decision-making found in multilateral treaty making processes, there are virtually no limits on what transnational actors can do to respond to climate change (Bulkeley et al. 2015: 164). This has massively increased participation in the global response to climate change, with the number of transnational initiatives now estimated to be in the thousands (Stevenson and Dryzek 2014: 87).

The development of transnational climate change governance has not only resulted in increased participation in climate change action, but also raised ambitions: many transnational initiatives include targets and timetables for reducing emissions of GHGs that go far beyond those agreed under the UNFCCC (Bulkeley and Newell 2010: 67). This is encouraging, because it creates a context where emissions can be lowered more substantially than under the multilateral regime. Furthermore, transnational actors can contribute to raising the ambitions of multilateral actors. Now that they have been invited to the annual Conference of the Parties (COPs), transnational actors have scope to raise their concerns and call for more stringent action. In fact, they are already doing so. At COP21 in Paris, members of the Climate Action Network (CAN) called on Parties to improve their pledges ahead of 2020 (ENB 2015: 13). Members of Climate Justice Now were slightly more forward, claiming that the celebration around the Paris Agreement illustrated 'palpable denial of multilateral actors' (ENB 2015: 13). Similarly, the reaction of cities, states and corporations in the United States (USA) to President Trump pulling out of the Paris Agreement was one of condemnation and defiance, with actors promising that they would meet the USA's GHG emissions targets defined in their Intended Nationally Determined Contribution (INDC), despite President Trump's decision to withdraw from the agreement (Tabuchi and Fountain 2017). This is indicative of transnational climate change actors' potential role in raising global ambitions.

In sum, existing research suggests that the nature of transnational climate change governance allows for increased participation, makes it possible to target emissions outside of the UNFCCC and raises ambitions for lowering emissions. This suggests that transnational climate change governance actors have the capacity to create a context within which emissions can be lowered, which creates space for the first part of Demand One to be enabled. In fact, recent studies focusing on the potential emissions reductions of transnational climate change actors by

2020 are quite optimistic about this capacity. For example, Sander et al. (2015: 467) found that commitments adopted by 238 leading city initiatives could reduce emissions by 2.8Gt of carbon dioxide-equivalent by 2020, and 13Gt by 2050. This is a significant reduction, equivalent to the emissions of all Organisation for Economic Co-operation and Development (OECD) countries in 2010. Similarly, a study released at COP21 in Paris focusing on a handful of transnational initiatives found their mitigation potential to be in the range of 2.5–4 billion tons of CO_2 by 2020, more than India emits in a year (Hale 2016: 13). Finally, a 2015 United Nations Environment Programme (UNEP) report found that transnational climate governance initiatives are now increasingly recognised as playing an important role in mitigating GHGs and bridging the global emissions gap (UNEP 2015: vi). Of course, these reports examine the potential of actors, rather than current emissions reductions, but UNEP (2015: vii) has noted that major projects in cities and regions are already delivering commitments that should result in emissions reductions even higher than was previously identified as possible.

The capacity to create a context where emissions can be kept in check, thereby enabling the first part of Demand One, will now be further explored with reference to two examples of transnational climate change governance: the C40, a public project that consists of a network of cities, and the Climate Group, a hybrid initiative that consists of public actors such as sub-state authorities as well as private actors such as corporations. Before these examples are explored, the chapter will briefly explain the importance of two types of actors within these projects: cities and corporations. Let us begin by considering the role of cities in the global climate change response. Cities are home to half the world's population, consume over two thirds of the world's energy and account for more than 70% of carbon emissions (Hoffman 2011: 104). It is estimated that by 2030, two thirds of the world's population will live in urban areas (Hoffman 2011: 104). It is therefore crucial that cities are involved in the process of lowering global emissions, because they account for a majority of emissions and encompass a majority of the global population, implying a high capacity to influence emission levels. Encouragingly, there is evidence that city efforts do not merely have capacity but have in fact already contributed to a reduction in global GHG emissions (McKendry 2016: 1357).

Such contributions are important not only in and of themselves, but also have knock-on effects. Cities who take ambitious action and pressure their federal governments to act on lowering emissions

(Pattberg 2010b: 152) can also disseminate knowledge. City networks can exchange best practices on issues ranging from energy-efficient buildings to water and waste treatment: they are key actors when it comes to disseminating applied knowledge and solutions to the challenge of climate change (Pattberg 2010b: 152). By spreading knowledge, cities could provide the momentum necessary to move toward decarbonisation (Hoffman 2011: 120). There is no one-size-fits-all way to reduce GHG emissions. Climate policies must be tailored to the concerns and needs of affected communities, which is why the city scale of action is so important (McKendry 2016: 1357). By implementing climate policies at political scales significantly smaller than the state, cities are able to demonstrate community-specific ways to act on climate change (McKendry 2016: 1357). Other cities can then be inspired or encouraged to act on climate change in a manner which suits their residents. For all of the reasons above, cities are a key part of creating a context where emissions can be lowered in order to protect the human right to health, and the C40 therefore presents a useful illustrative example of transnational climate change governance.

Similarly, corporations play a critical role in the production of global emissions. The private sector accounts for more than one third of energy consumed worldwide and corporations often emit significantly more GHGs than major cities (Bulkeley and Newell 2010: 87). A Greenpeace study, for example, found that Shell emits more than Saudi Arabia, Amoco more than Canada, Mobil more than Australia and British Petroleum, Exxon and Texaco more than France, Spain and the Netherlands combined (Bulkeley and Newell 2010: 2). This indicates that corporations produce a high percentage of global emissions and must necessarily be part of the global effort to reduce emissions due to their capacity to influence the global level of GHGs. Beyond simply lowering emissions, groups of corporations can effectively institutionalise new norms at the transnational level; for example, the norm to disclose corporate carbon emissions (Pattberg 2010b: 159). These emerging norms are expected to motivate and facilitate meaningful dialogue among business actors, investors and the wider public to induce corporate responses to climate change (Pattberg 2010b: 160). Furthermore, these actors have a role to play in the financing of the climate change response. The United Nations estimates that 98% of global investment and financial flows required to tackle climate change will need to come from the private sector, because the private sector develops and disseminates most of the world's technology (Bulkeley and Newell 2010: 88). So, although corporations are no doubt part of the reason climate change is occurring due to their high emissions, it seems that these actors can also be a crucial part of the solution. It is for this reason that projects such as the Climate Group are a key part of creating a context within which emissions can be lowered enough to protect the right to health of future generations.

The C40 and the Climate Group will now be briefly discussed in order to further the argument that transnational climate change governance actors have the capacity to enable the first part of Demand One by creating a context where this is possible. The C40 is a network of the world's largest cities which aims to share best practices and develop collaborative initiatives on city-specific issues in order to make implementing the global response to climate change more feasible (Hoffman 2011: 95). Through this, the C40 (C40 2017b) promises to have 'a meaningful global impact in reducing both greenhouse gas emissions and climate risks'. Importantly, only twenty of the ninety C40 cities are held to account for emissions reductions under the Kyoto Protocol. This is quite promising in terms of creating a context for lowering emissions because if the seventy-two cities not held to account under the Kyoto Protocol lower their emissions, this will add to multilateral governance efforts (which account for 15% of global emissions until 2020). In addition to holding new actors to account, the C40 also encompasses a large part of the global population: representing over 600 million people, which amounts to one in twelve individuals worldwide (C40 2017b). The C40 therefore has the capacity to make a very real impact on climate change mitigation by creating a broad network, or context, within which emissions can be reduced.

It seems the creation of such a context is well under way. Cities who are members of the C40 are three times more likely to act on climate change (C40 2017a). Since 2011, actions being delivered at a city-wide scale within the C40 network increased by 260% (C40 2015). The C40 currently has over 10,000 climate change initiatives in place, falling into several categories: adaptation and water, energy, finance and economic development, measuring and planning, solid waste management, sustainable communities and transportation (C40 2017a). This is a wide distribution of activities, which is promising because the C40 has the potential to help cities reduce their emissions in multiple ways. Indeed, the C40's goals of information sharing and providing concrete examples of measures that reduce emissions has proven to be an effective tool for motivating cities to take action they likely otherwise would not have (Bulkeley et al. 2015: 171).

However, and perhaps unsurprisingly given the discussion above, there is a lack of data on the C40's efforts. Although the C40 released a report in 2014 which claims to 'provide compelling evidence of the importance of the C40 network', this report does not make mention of emissions reductions (C40 2014: 3). There are also no independent reports verifying their emissions reduction efforts. Nevertheless, it is clear that the C40 has the capacity to create a context within which global emissions can be lowered, because it includes a broad range of cities, encompasses a large part of the global population and has over 10,000 current projects in place, addresses a multitude of issue areas and ensures that cities can pursue emissions. Therefore, although it is not possible to definitively claim that the C40 has had a substantial effect on global emissions levels, the initiative quite clearly creates a context under which this is possible.

The Climate Group is another useful example of the capacity of transnational governance actors to create a context where emissions can be lowered, because it aims to 'deliver a world of net zero greenhouse gas emissions and greater prosperity for all' (Climate Group 2017a). More specifically, the Climate Group aims to create a clean revolution through the rapid scale-up of low-carbon energy and technology (Climate Group 2017a). The Climate Group is made up of over one hundred major corporations, sub-national governments and international institutions (Climate Group 2017b). These corporations and institutions are spread across the globe: corporate members range from Ikea (Sweden) to Dell (America) and Taobao (China), and public members include California, South Australia, Rio de Janeiro, Mumbai, Kolkata and Toronto (Climate Group 2017a). Importantly, out of the ninety cities, states and provinces participating in the Climate Group, only eighteen are held to account for lowering emissions under the Kyoto Protocol. If emissions are lowered by the seventy-two cities that are not held to account under the Kyoto Protocol, this could contribute to the lowering of global emissions, because this will add to multilateral governance efforts.

In addition, as was explained in Chapter 5, corporations are not held to any direct account under the Kyoto Protocol. Instead, states are held to account and can then set targets for corporations within their own borders. If corporations that exist outside of these states reduce their emissions because of the Climate Group, this will add to multilateral mitigation efforts. In addition to representing a substantial number of cities and corporations not held to account under the Kyoto Protocol, members of the Climate Group represent a significant amount of wealth and global population. The Climate Group claims that the combined revenue of its corporate members is in excess of $1 trillion USD, while its city and regional partners represent almost half a billion people (Climate Group 2017b). This is promising in terms of possible financial contributions to the climate change cause, and for lowering global emissions, because the Climate Group encompasses a substantial part of the population and has wealthy corporate backers that can help implement its initiatives. This creates a context within which global emissions can be lowered significantly.

In order to promote the scale-up of clean technology, the actors in the Climate Group pilot solutions that can be replicated worldwide (Climate Group 2017a). For example, the LED Lighting Project, in place in major cities such as London, New York, Hong Kong and Mumbai, has shown that switching to LED lights in cities can present energy savings as high as 80% (Climate Group 2012: 24). This is quite promising, especially if this kind of energy saving can be applied to all cities across the world. Street lighting accounts for 6% of global emissions levels, which is the equivalent of GHG emissions from 70% of the world's passenger vehicles. If the LED Lighting Project is applied globally it would contribute to lowering the 6% of global emissions associated with street light down to 1.2%. Through projects such as the LED Light Project, the Climate Group aims to 'break the climate deadlock' and advocate for stringent global action by demonstrating the availability of solutions (Hoffman 2011: 85). This indicates that the Climate Group is attempting to create a context within which emissions can be lowered, by demonstrating solutions that can make a difference to emissions levels. Unfortunately, there is not much concrete evidence available on whether the Climate Group is able to live up to its ambitions, which is perhaps unsurprising given that there is a lack of evidence for this across the board of transnational climate change governance projects. Nevertheless, it is evident from the above that the Climate Group has the capacity to create a context within which emissions can be lowered.

The discussion and examples above illustrate that transnational climate change actors are working towards creating a context which allows for increased participation and increased ambition, thereby raising expectations about what it is possible to achieve. Some initiatives, such as the C40 and Climate Group, explicitly aim to lower emissions, further indicating that transnational climate change governance actors have the capacity to lower global emissions. Of course, the evidence for emissions reductions is severely lacking, calling into question to what extent such a context is being created. Nevertheless, although it is not possible to claim that transnational climate change governance actors fully enable the first part of Demand One, because this cannot be definitively proven, it is clear that these actors have great potential. In terms of the climate justice hierarchy, transnational climate change governance actors are therefore on rung three of the four-point hierarchy in terms of the first part of Demand One:

> 3. Actors in the institution have promised to begin working on enabling the demand of justice in the future – no policy has been adopted, but there is the potential for the creation of policy in order to consistently work towards enabling the demand of justice.

Although it cannot be said that all actors within transnational climate change governance have promised to begin working on enabling the first part of Demand One in the future, because this is impossible to claim due to vast number of projects in existence, the discussion has served to tentatively illustrate the potential for policy which enables the first part of Demand One. The potential for policy lies in the fact that there is a context under which policy for lowering emissions can be created, and that some initiatives, like the C40 and Climate Group, are already pursuing such policies. Unfortunately, there is not enough evidence to suggest that these policies are consistently enabling the first part of Demand One. For this reason, transnational climate change governance actors remain on the third rung of the hierarchy.

Finally, although there is potential for policy, there is also room for pessimism. So far, transnational climate change governance has not delivered anything remotely close to matching the scale of the climate change problem (Stevenson and Dryzek 2014: 3). While it may be too soon to tell how these actors will fare in future, transnational climate change governance remains a long way from achieving significant emissions reductions (Bulkeley 2015: 171). This has led some critics to claim that transnational climate change governance actors are merely pursuing their own interests, rather than making fundamental changes to the global response to climate change. Whether this is a fair judgement will be further discussed in the final part of the chapter, which will summarise what has been found. For now, it is worth reiterating that determining the effectiveness of transnational climate change governance in terms of emissions reductions is very problematic, and could lead to a premature dismissal of governance initiatives as a key part of the global response to climate change (Bulkeley et al. 2015: 161). For this reason, the chapter now moves on to exploring whether actors involved in transnational climate change governance enable the second part of Demand One, as well as Demands Two and Three, in order to get a better sense of the overall role of transnational climate change governance actors can play in enabling a condition of climate justice.

Adaptation

The second part of Demand One calls for the prioritisation of adaptation alongside mitigation. Assessing to what extent adaptation has been prioritised is not a straightforward process. One important area of focus is the extent to which transnational actors have been working towards raising the funds required for adaptation, which are estimated to be $125 billion USD by 2050 (Hof et al. 2010: 252). However, adaptation is not merely a matter of financial cost, it is also about sharing technology and ensuring that countries vulnerable to climate change are assisted in preparing for climate change. The assessment will therefore focus on finance, technology and assistance. In this way, the chapter endeavours to take the complexities of adaptation into account. Of course, not all complexities can be captured in this short discussion. Instead, the chapter provides an overview of transnational action on adaptation that can be built upon in future research.

Unfortunately, existing research on transnational climate change governance reveals that transnational actors appear to strongly favour mitigation over adaptation, much like multilateral actors. In Hoffman's study of fifty-eight projects, for example, 69% focus exclusively on mitigation, 28% pursue a mix of mitigation and adaptation efforts and only 3% focus on adaptation alone (Hoffman 2011: 40). Similarly, Bulkeley et al.'s 2015 study found that 75% of sixty projects focus on mitigation alone, 22% focus on both mitigation and adaptation and only 3% focus solely on adaptation (Bulkeley et al. 2015: 130). Interestingly, when these scholars examined projects involving members from less developed countries, the numbers varied only slightly: 64% of projects focused exclusively on mitigation, 22% on both mitigation and adaptation and 4% on adaptation alone (Bulkeley et al. 2015: 130). This indicates that even when less developed country actors are part of transnational projects, less developed country concerns are not represented. This may be down to the exclusion of countries from regions that are most in need of assistance with adaptation. The exclusion of these countries will be further discussed when assessing to what extent transnational actors enable Demand Two of justice. For now, existing research demonstrates that a context where adaptation is prioritised alongside mitigation has not yet been created. Although it is encouraging that around a quarter of projects pursue both mitigation and adaptation, these numbers are dwarfed by around two thirds of projects pursuing mitigation only. Most alarming is the fact that so few transnational actors are pursuing adaptation alone – around 3%, according to existing research. Adaptation is clearly not being prioritised alongside mitigation, which means the second part of Demand One is not being enabled by transnational climate change governance actors.

When considering specific adaptation measures, such as financing and technology transfer, a similar picture emerges. Out of the 1,073 projects at the New York Climate Summit in 2015, just one, the New York Declaration on Forests, included an explicit financial pledge of $450 million USD, and only two projects included specified financial commitments at all (Hsu et al. 2015: 501). This is alarming, considering that paying for adaptation, and quickly, is required to protect the human right to health. On the continent of Africa alone, the annual adaptation costs grow by 10% each year (Abeysinghe and Huq 2016: 195). Funds must become available, and the sooner the better, to prevent spiralling costs and significant impacts on human health for the most vulnerable countries. However, there might be some optimism when looking to the future. The 2014 Global Investor Statement on Climate Change, for example, was signed by 348 investors representing more than $24 trillion USD in assets (UNEP 2015: 20). These actors have pledged to 'work with policy makers to support and inform their efforts to develop and implement policy measures that encourage capital deployment at scale to finance the transition to a low carbon economy and encourage investment in climate change adaptation' (UNEP 2015: 20). They have also pledged to invest 'in areas such as renewable energy, energy efficiency and climate change adaptation' (UNEP 2015: 20). This indicates that transnational actors are willing to contribute to the financial costs of adaptation, as well as assisting with developing technology that could be transferred to less developed countries to assist them with adaptation. This may be indicative of a broader change in transnational climate change governance, especially considering that the most recent study on the subject focused on initiatives which ran between October 2008 and March 2010 (Bulkeley et al. 2015: 23). The attitude towards adaptation may have shifted significantly since this time.

It is also important to point out that the statistics on how many transnational initiatives pursue adaptation are potentially misleading. Adaptation efforts are often not couched explicitly in climate governance terms, because they focus on local concerns, which means they may have been missed by researchers (Hoffman 2011: 40). Adaptation-focused initiatives could also be missed by researchers because they fall into the 'fuzzy area' of sustainable development, or are focused on specific problems like flood control, soil erosion, land management, emergency response or health (Hoffman 2011: 40). In this sense, the small number of adaptation-focused projects found in previous research may paint an inaccurate picture of the climate change governance landscape. Perhaps future research projects will need to change their methodologies in order to capture the number of adaptation-focused projects more accurately. Finally, there is room for optimism in the sense that adaptation has only recently emerged as a key feature of transnational climate governance (Hoffman 2011: 40). There are hopes that adaptation will become an increasingly important component of the global response to climate change in the future, as transnational climate change governance actors come to see adaptation as a key part of climate change governance.

To further illustrate why there is room for optimism in terms of the second part of Demand One, the chapter will now provide two examples of transnational climate change governance initiatives focusing exclusively on adaptation: The Asian Cities Climate Change Resilience Network, and 100 Resilient Cities. The Asian Cities Climate Change Resilience Network is a city-focused public initiative, and 100 Resilient Cities is a hybrid initiative that focuses on cities but has corporate backers and non-governmental organisation (NGO) members. These two city-focused examples are useful because cities play an important role in furthering adaptation. As was explained above, cities are home to the majority of individuals of the world. If cities pursue adaptation measures, they are capable of protecting the right to health of a significant number of people. In addition, as with mitigation measures, cities can provide examples of good practice of adaptation for other cities by demonstrating how local adaptation measures can be financed and implemented.

The dissemination of such knowledge is crucial for ensuring adaptation is prioritised, because it makes it possible for other cities to understand how to act on climate change and prepare themselves effectively. Furthermore, localised climate adaptation efforts may enable the broad-based participation, articulation of perceptions of vulnerability and risk, and analysis of differentiated impacts among different groups and communities that proponents of climate justice call for, rather than run the risk of exacerbating existing injustices in the name of the urgency of rapid, global GHG reductions (McKendry 2016: 1357). In other words, local adaptation measures allow those affected to have a direct say on how they are protected. This is often difficult to achieve in multilateral climate change governance. For all of the reasons above, cities play a crucial role in adaptation, and the Asian Cities Climate Change Resilience Network and 100 Resilient Cities present useful illustrative examples of transnational climate change governance. Let us first turn to the Asian Cities Climate Change Resilience Network.

The Asian Cities Climate Change Resilience Network (ACCCRN) aims to strengthen the capacity of over fifty rapidly urbanising cities in Bangladesh, India, Indonesia, the Philippines, Thailand and Vietnam to 'survive, adapt, and transform in the face of climate-related stress and shocks' (ACCCRN 2017a). Further to this, the ACCCRN aims to 'enable poor, marginalised, and otherwise vulnerable people in Asia's emerging cities to be included and supported in the systems and processes driving urbanisation and emerging resilience-building measures' (ACCCRN 2017b). By doing so, the ACCCRN hopes to give marginalised communities an increasingly prominent voice in their cities' future (ACCCRN 2017b). In this sense, the ACCCRN is not only creating a context where adaptation is prioritised, but also where those affected can be part of decision-making on what adaptation looks like. Considering that multilateral governance actors have had trouble including these voices, this is a significant contribution. It is, however, important to point out that although Bangladesh, India, Indonesia, the Philippines, Thailand and Vietnam are all considered wealthier less developed countries under the definition of the UNFCCC (Non-Annex), only Bangladesh is part of the group of Least Developed Countries, which are the most vulnerable to climate change effects. This means that the ACCCRN primarily assists less developed countries, rather than those most vulnerable to climate change, suggesting that the most marginalised voices are not being represented under this particular initiative.

The ACCCRN operates in five countries (Bangladesh, India, Indonesia, the Philippines, Thailand and Vietnam) and has over 2,000 members of staff, including practitioners, academics, government officials and professionals (ACCCRN 2017c). The scope for prioritising adaptation is therefore quite wide, encompassing a population of over two and half billion people. The ACCCRN has dozens of current projects that help to illustrate their potential to prioritise adaptation. Let us consider two such projects. The first is the 'Indoor Temperature Comfort' initiative in Indore, India, which seeks to promote indoor thermal comfort through no/low energy options for urban residents (ACCCRN 2017d). Considering that heat stress poses a direct threat to human health because it can cause or aggravate cardiovascular and respiratory disease, particularly among elderly people (WHO 2017), this is an important adaptation project that could contribute to the protection of the human right to health.

A second example of the ACCCRN's potential is the long-term flood resilience project currently taking place in Hat Yai, Thailand. This project aims to build a participatory platform for the development of strong public–private-citizen partnerships for the planning and implementation of integrated flood risk reduction measures (ACCCRN 2017d). The expected outcome is the formulation of integrated flood risk reduction plans that will be planned, developed and endorsed by public–private-citizen actors as options for implementation and scaling up. This is a good example of ACCCRN attempting to include the voices of those who will be affected by climate change. Furthermore, considering that flooding can affect clean water supplies and thereby spread disease, this is an important project in terms of protecting the human right to health (IPCC 2014a: 15). Overall, then, it seems the ACCCRN is creating a context where adaptation is prioritised and the human right to health can be protected. This is encouraging, given the so far pessimistic research about the potential of transnational climate change governance to prioritise adaptation.

100 Resilient Cities (100RC) is another good example of a transnational climate change governance initiative that prioritises adaptation. Set up by the Rockefeller Foundation, 100RC is 'dedicated to helping cities around the world become more resilient to the physical, social and economic challenges that are a growing part of the 21st century' (100RC 2017a). Members of 100RC support the 'adoption and incorporation of a view of resilience that includes not just the shocks – earthquakes, fires, floods, and so on – but also the stresses

that weaken the fabric of a city on a day to day or cyclical basis' (100RC 2017a). In this sense, 100RC has quite a broad definition of adaptation, and their work has a wide remit within cities. The initiative is currently made up of 163 cities, with plans to expand in future (100RC 2017b). Out of these 163 cities, forty-five (28%) are in less developed countries according to the UNFCCC (Non-Annex) definition, and five (3%) are in Least Developed Countries: Addis Ababa, Ethiopia; Paynesville, Liberia; Mandalay, Myanmar; Kigali, Rwanda; Dakar, Senegal. Considering that developing countries need more adaptation assistance than developed countries, 28% vs 69% is not a very high ratio. Perhaps more troublingly, the fact that only 3% of cities are from the most vulnerable countries suggests that 100RC has a long way to go to protect those most in need.

Nevertheless, 100RC has an impressive remit of action on adaptation. Member cities are offered the following (100RC 2017a):

1. Financial and logistical guidance for establishing an innovative new position in city government, a Chief Resilience Officer, who will lead the city's resilience efforts.
2. Expert support for development of a robust Resilience Strategy.
3. Access to solutions, service providers, and partners from the private, public and NGO sectors who can help them develop and implement their Resilience Strategies.
4. Membership of a global network of member cities who can learn from and help each other.

From the list above, it is clear that the 100RC goes far beyond mere financial assistance and focuses instead on knowledge transfer and capacity building. Of course, financial assistance is part of their plans (each city is likely to receive an excess of $1 million USD), but the primary objective of 100RC is to teach cities how to adapt to climate change. This is very important, because adaptation is a complex matter which requires much more than technological or financial transfer – it requires specialised local knowledge for long-term implementation and effectiveness. In this sense, 100RC is very much capable of creating a context where adaptation can be prioritised. Indeed, such a context is well under way to being created. Since its inception in 2013, 100RC has implemented over 1,600 concrete actions and initiatives and currently has over 13,000 members of the community working on urban resilience in their cities

(100RC 2017c). They have delivered 10,500 hours of training and part-nered with 2,500 local community groups (100RC 2017c). This scale is highly impressive given how young the project is (four years at the time of writing). Involving local actors at such a scale is particularly remark-able, considering how much the multilateral regime has struggled to do so. Much like the ACCCRN, 100RC aims to 'ensure citizen voice and ownership of the resilience agenda'. It seems they are well on the way to creating a context where adaptation is not only prioritised but led by those who will be affected by climate change. This is very much in line with the second part of Demand One. 100RC is therefore an excellent example of a transnational climate change governance initiative, which, although rare, demonstrates the potential of transnational actors to cre-ate a context where adaptation is prioritised alongside mitigation.

Although only very few initiatives, including the ACCCRN and 100RC, explicitly focus on adaptation, there is some room for optimism that this will change over time, considering the age of the most recent studies and the fact that existing methodologies may have excluded initiatives. In addition, there have been promises made, specifically through the 2014 Global Investor Statement on Climate Change, to raise money for adaptation and enable technology transfer and assis-tance. Therefore, although there are remaining problems, including an overwhelming majority of projects focusing on mitigation, and although it is not possible to claim that transnational climate change governance actors fully enable the second part of Demand One, it is clear that these actors have potential to do so in future. In terms of the climate justice hierarchy, transnational climate change governance actors are therefore on rung three of the four-point hierarchy in terms of the second part of Demand One:

> 3. Actors in the institution have promised to begin working on enabling the demand of justice in the future – no policy has been adopted, but there is the potential for the creation of policy in order to consistently work towards enabling the demand of justice.

While it cannot be said that all actors within transnational climate change governance have promised to begin working on enabling the second part of Demand One in the future, because this is impossible to claim due to the vast number of projects in existence, the discussion has served to illustrate the potential for policy. This potential lies in the fact that there is a context under which policy that prioritises adaptation can be created, and that some initiatives, like the ACCCRN and 100RC, are already pursuing such policies. Unfortunately, there is not enough evidence to suggest that these policies are consistently enabling the second part of Demand One. For this reason, transnational climate change governance actors remain on the third rung of the hierarchy. Finally, although there is room for optimism, change is not guaranteed. There is currently clearly not enough focus on adaptation, and even when there is, the most vulnerable are not being prioritised (Sander et al. 2015: 467). Transnational climate change actors must become much more active and focused on adaptation if the second part of Demand One has any hopes of being realised. This will be further explored in the final part of the chapter. For now, the chapter turns to Demand Two.

DEMAND TWO – THE RESPONSIBILITIES OF STATES

Demand Two asserts that the distribution of benefits and burdens in global climate change action must be based on the PATP model.[1] Enabling Demand Two therefore requires that the responsibility to contribute finances and lower emissions is based on both emissions levels and wealth levels. The chapter will investigate to what extent transnational actors are enabling this model of responsibility by examining existing research on the subject and exploring two key examples: the Carbon Sequestration Leadership Forum and the Asia-Pacific Partnership on Climate. In addition, given that multilateral actors have struggled to meaningfully include less developed countries in decision-making procedures, the chapter will also assess whether transnational actors have fared any better on this matter. This is an important aspect to consider, because as Chapter 3 explained, the PATP model is not enough to guarantee the fair treatment of less developed countries. Ideally, they would be included in decision-making processes, to ensure that their voices are heard. The discussion below will therefore also explore to what extent less developed country members have been part of decision-making procedures within transnational climate change governance.

Let us first consider the allocation of responsibility between states under transnational climate change governance. Unfortunately, it is difficult to claim with any certainty that transnational climate change governance actors create a context where states are held to account in line with the PATP model. This is for two reasons. The first is that states are not the primary actors within transnational climate change governance, and the second is that even when states are the primary actors within a project, they are not often held to account more than voluntarily. In Bulkeley et al.'s sample of sixty initiatives, for example, only 15% were initiated by national government actors (Bulkeley et al. 2015: 26). In addition, in their sample of public projects, only 22% had national government actors, compared to 38% that had regional actors, 31% that had local actors and 9% that had multiple public actors (Bulkeley et al. 2015: 82). This is indicative of the low number of state actors within transnational climate change governance, which is problematic in terms of enabling the PATP, because if states do not participate, they cannot be held to account. Furthermore, even when states do participate in initiatives, it is rare for these actors to be held to account more than voluntarily. For example, in Bulkeley et al.'s analysis, they discovered that capacity building (88%) and information sharing (93%) are the most common functions of transnational initiatives (Bulkeley et al. 2015: 27). They also found that a significant proportion of initiatives undertake direct forms of action (60%) and involve setting some form of target for their constituents (60%) (Bulkeley et al. 2015: 27). However, initiatives that set mandatory rules (23%) are relatively rare (Bulkeley et al. 2015: 28).

If states are not held to account more than voluntarily, it is questionable whether the PATP model of responsibility is being enabled by transnational climate change governance actors. Nevertheless, in Bulkeley et al.'s sample of sixty initiatives, a majority have some form of 'soft' or 'self' regulation. This indicates that transnational governance initiatives have ways to hold actors to account. These strategies include: maintaining a registry of members (77%), asking participants to sign a Memorandum of Understanding (35%), charging a membership fee (37%) or requiring compulsory action (30%) (Bulkeley et al. 2015: 34). In fact, 90% of initiatives undertake some form of function that goes beyond information sharing or capacity building to include certification or target setting, which indicates actors are being held to account, even if this is through soft, voluntary measures (Bulkeley et al. 2015: 35). Therefore, although it cannot be said that actors are held to account by a legal treaty such as the Kyoto Protocol, transnational climate change governance actors have created a context within which actors can be held to account through soft regulation. Two examples of transnational climate change governance initiatives will now be examined to illustrate how this soft regulation works in practice. These two examples, the Carbon Sequestration Leadership Forum and the Asia

Pacific Partnership on Clean Development and Climate, are both public initiatives that exclusively have states as members. The two examples illustrate that transnational governance actors have created a context within which rich and high-emitting states can act on their responsibilities and hold one another to account, albeit on a voluntary basis.

The Carbon Sequestration Leadership Forum (CSLF) is focused on the 'development of improved cost-effective technologies for the separation and capture of carbon dioxide for its transport and long-term safe storage' (CSLF 2017a). The mission of the CSLF is to 'facilitate the development and deployment of such technologies via collaborative efforts that address key technical, economic, and environmental obstacles' (CSLF 2017a). Membership is open to national governmental entities that are significant producers and/or users of fossil fuels and that can invest resources in research, development and demonstration activities in carbon capture and storage technologies (CSLF 2017a). This indicates that the CSLF specifically targets high-emitting and/or wealthy countries that can contribute to funding new technologies. Importantly, the CSLF includes members from less developed countries that are widely regarded as significant in terms of their contributions to GHG emissions and as critical actors in the international climate change regime. These are richer less developed countries: Brazil, India, China, South Africa, as well as Russia and Mexico (BRICSAM). The CSLF also includes the USA, which never ratified the Kyoto Protocol and is now pulling out of the Paris Agreement. In addition, it includes Canada, Japan, Russia and New Zealand, countries who have refused to participate in the second round of the Kyoto Protocol. In this way, CSLF encompasses actors which not currently held to account under the Kyoto Protocol. Overall, the CSLF has twenty-six members (twenty-five plus the European Commission).

The PATP calls for high-emitting and/or rich countries to contribute to the climate change solution, by lowering emissions and/or financing projects that seek to address the climate change problem. The CSLF is in line with these demands, because it allows high-emitting and wealthy states to act on climate change, according to their responsibilities under the PATP model. Of course, these actors are not held to account on more than a voluntary basis, as is typical within transnational climate change governance. Although the CSLF has a Charter, this is 'does not create any legally binding obligations between or among its members' (Bulkeley et al. 2015: 32). Furthermore, the Charter of the CSLF specifies that 'a member may withdraw from membership in the CSLF by giving

90 days advance written notice to the Secretariat', indicating that membership is not legally binding (CSLF 2017b). Nevertheless, the CSLF provides a context within which these actors can act on their responsibilities and hold one another to account for action on climate change, even if this is on a voluntary basis. This is indicative of the fact that transnational climate change governance actors are capable of creating a context within which rich and/or high-emitting countries can act on, and be held accountable for, their responsibilities under the PATP model, thereby enabling Demand Two.

The Asia-Pacific Partnership on Climate (APP) is another example of a transnational climate change initiative that allows high-emitting/rich countries to act on their responsibilities specified under the PATP model. Although the APP concluded its joint work in 2011, it is a frequently used as an example within the climate change governance literature: Hoffman (2011: 8), Bulkeley et al. (2015: 32) and Abbott (2012: 575) all discuss the APP in their research. It is therefore worth briefly discussing this example, because this will allow the current chapter to speak to existing literature. The APP accounted for 48% of the world's GHG emissions, 48% of global energy production, 49% of global GDP and 45% of the global population, making it a significant transnational climate change governance actor, which is perhaps why it is commonly discussed (Bäckstrand 2008: 92). The APP was a partnership between the governments of Australia, Canada, China, India, Japan, South Korea and the United States, and ran from 2005–11. Besides Australia, none of the APP member states are currently held to account for lowering emissions or financial contributions under the Kyoto Protocol. This is important, because it illustrates that the APP, like the CSLF above, created a context under which rich/high-emitting countries not held to account under the Kyoto Protocol can act on their responsibilities, thereby enabling Demand Two.

The APP aimed 'to promote and create an enabling environment for the development, diffusion, deployment and transfer of existing and emerging cost-effective, cleaner technologies and practices, through concrete and substantial cooperation' (Bulkeley et al. 2015: 32). This indicates that the APP was concerned with investing in technologies which lower GHG emissions, similarly to the CSLF, and members of the APP contributed financially in order to ultimately lower global emissions. Over the course of its existence the member states of the APP were involved in hundreds of projects to fulfil this aim. As one example, the APP oversaw the construction of high-performance buildings in China, which aimed to reduce energy use and GHG emissions. One of these buildings was the Olympic Village Micro-Energy Building, which housed 17,000 athletes during the 2008 Olympics. Through projects such as these, the APP allowed rich and high-emitting actors who are responsible under the PATP model to act on their responsibilities.

However, members of the APP were not held to account beyond their voluntary commitment. According to the APP charter, the member states were operating under a 'voluntary, non-legally binding framework' and partners were allowed to terminate their membership 'upon written notice 90 days prior to the anticipated termination' (APP 2014). In addition, although the APP enabled states to act on their responsibilities, the APP reflected a preference for an emphasis on reducing emissions intensity rather than cuts in carbon per se (Bulkeley et al. 2015: 75). Furthermore, the APP had a strong emphasis on clean coal technologies, which reflects the energy sector profiles of some of the countries involved (Bulkeley et al. 2015: 75). This indicates that the APP presented a 'business as usual' approach, rather than a radical departure from existing energy production. This is a wider problem within transnational climate change governance that will be further discussed in the final part of the chapter.

For now, the discussion above has aimed to demonstrate that transnational climate change governance actors have the capacity to create a context within which rich/high-emitting states can act on their responsibilities, albeit voluntarily. In addition, the two examples above served to illustrate that there are existing initiatives that involve rich and high-emitting states acting to reduce emissions and fund climate friendly technologies, thereby setting a context where the PATP model of responsibility is followed and Demand Two is enabled. For this reason, the actors involved in transnational climate change governance can tentatively be said to reside on rung three of the four-point justice hierarchy in terms of enabling Demand Two:

3. Actors in the institution have promised to begin working on enabling the demand of justice in the future – no policy has been adopted, but there is the potential for the creation of policy in order to consistently work towards enabling the demand of justice.

Although it cannot be said that all actors within transnational climate change governance have promised to begin working on enabling

Demand Two in the future, because this is impossible to claim due to the vast number of projects in existence, the discussion has served to illustrate the potential for policy, which lies in the fact that there is a context under which the PATP could be inscribed into policy. The Carbon Sequestration Leadership Forum (CSLF) and the Asia-Pacific Partnership on Climate (APP), both of whom enable states to act on their responsibilities, are indicative of this potential. Nevertheless, it cannot be said that transnational climate change governance has policies in place that consistently enable the PATP, because in the rare cases where states make up actors within these processes of governance, they are not held to account more than voluntarily. For this reason, transnational climate change governance actors remain on the third rung of the hierarchy in terms of Demand Two.

Let us now briefly examine whether transnational actors fare any better than multilateral actors in terms of including less developed countries in decision-making procedures. Unfortunately, the consensus in transnational climate change governance research is that less developed countries are generally marginalised. In Bulkeley et al.'s sample of sixty projects, for example, the majority of projects (87%) were initiated by actors in developed countries (Bulkeley et al. 2015: 32). Pattberg (2010b: 151) has made a similar observation, noting that transnational climate change governance seems to favour actors from a certain area of the world: namely Western, developed nations, primarily the USA, Canada and Australia. Furthermore, in Hoffman's sample of fifty-eight projects, only 12% were initiated by a combination of actors in developed and less developed countries, and 88% were initiated by developed country actors (Hoffman 2011: 30). This lack of inclusion is problematic – these countries cannot contribute to decision-making if they are not present, and the decisions made in their absence may not reflect their concerns. This may be why a 'bulk of less developed countries' continue to support multilateral climate change governance (Biermann et al. 2010: 31).

Nevertheless, it is worth noting that transnational climate change responses are expected to broaden participation and increase the inclusion of otherwise marginalised actors (Pattberg 2010a: 279). In this way, transnational climate change governance actors may create a context within which less developed countries are included in decision-making over time. Interestingly, the newest studies indicate that transnational climate change governance is moving in this direction. Bulkeley et al. (2015: 32), for example, found that 77% of their sixty initiatives have at least one actor from a less developed country, and 67% have at least two. Importantly, out of the forty-six projects that include less developed country members, 50% involve non-BRICS participants, indicating that it is not only the richer less developed countries that are participating (Bulkeley et al. 2015: 32). If transnational climate change governance were predominantly driven by what was occurring under the UNFCCC, BRICS countries would be expected to be the primary participants because of their growing role in multilateral governance, but this appears not to be the case (Bulkeley et al. 2015: 32). This indicates that transnational climate change actors are creating a context where less developed countries can be part of decision-making procedures more so than at the multilateral level.

However, only 7% of Bulkeley et al.'s sample solely involves participants from developing countries (Bulkeley et al. 2015: 120). In addition, these scholars found a significant regional variation to participation, with the regions of Sub-Saharan Africa (SSA), Oceania and the Middle East and North Africa (MENA) remaining particularly underrepresented (Bulkeley et al. 2015: 123). Bulkeley et al.'s findings indicate that North America and Europe are at the core of the transnational governance world, closely followed by Asia. Overall, these regional patterns mimic closely the way that diplomats from large parts of the developing world have less capacity to shape the multilateral regime (Bulkeley et al. 2015: 125). Importantly, underrepresentation can directly affect which interests are taken into account. For example, for those actors most marginalised (SSA and Oceania), adaptation is a key concern. As was discussed above, transnational climate change governance actors appear to favour mitigation over adaptation – perhaps as a direct result of the exclusion of the most vulnerable. And yet, there is some room for optimism because of the high participation by less developed countries found in Bulkeley et al.'s study, especially as this finding was unexpected for the scholars. Their research indicates that transnational climate change governance actors create a context where less developed country participation is quite diverse. So, although there is a long way to go in terms of ensuring that less developed countries are included in decision-making procedures, there seems to be some progress being made in transnational climate change governance. Nevertheless, it is very clear that a version of the traditional North–South divide is being replicated in transnational climate change governance, with more participation and more leadership from developed than from developing countries

(Hale 2016: 20). This will be further discussed in the final part of the chapter. For now, the chapter turns to the third and final demand of climate justice.

DEMAND THREE – THE RESPONSIBILITIES OF NON-STATE ACTORS

Demand Three asserts that all capable actors, including individuals, firms, sub-state entities and international institutions, irrespective of the country in which they live or exist, must be held responsible for lowering emissions and/or contributing to adaptation efforts, in line with their respective capabilities. To enable Demand Three, transnational actors must create a context where non-state actors are held to account for their climate-related responsibilities. To assess whether this has been the case, the chapter will analyse existing climate change governance research and examine two key examples: the Verified Carbon Standard and the Regional Greenhouse Gas Initiative.

Encouragingly, recent research on the transnational climate change governance is indicative of their capacity to create a context within which responsible non-state actors can respond to climate change. For example, Bulkeley et al.'s (2015: 1) findings indicate that it is becoming increasingly common for subnational governments, non-governmental organisations, businesses and individuals to take responsibility into their own hands and experiment with bold new approaches to the governance of climate change. In fact, their research revealed a clear increase in participation of non-state actors over the past decade (Bulkeley et al. 2015: 68). A study released at COP21 in Paris found that there are now climate commitments from over 7,000 cities from more than ninety-nine countries, with a combined population of 794 million (11% of the global population and around 32% of global GDP), as well as close to 5,000 companies from over eighty-eight countries representing over $38 trillion USD in revenue (Hale 2016: 13). Research suggests that these actors are not merely participating: transnational climate change governance processes actively exercise authority over non-state actors, including individuals, companies and even intergovernmental organisations (Bulkeley et al. 2015: 3). This indicates that transnational climate change governance actors have the capacity to create a context within which non-state actors can not only act on their responsibilities but be guided by an authoritative transnational agent in their attempts to do so.

It is also promising that transnational climate change governance is not only widening participation of non-state actors, but resulting in new kinds of climate change responses. Recent research suggests that transnational climate change governance processes provide scope and space for action which is broader than the space provided by the UNFCCC (Bulkeley et al. 2015: 165). Transnational climate change responses do not face the same constraints that affect decision-making at the multilateral level: if the interest and money is there, action can be taken without having to negotiate the consent of 180 states. In this sense, transnational climate change governance processes can circumvent negotiation stalemates among countries that might have been caused by the attempt of finding universal agreement (Biermann et al. 2010: 30). This, in turn, may make it easier to broaden the coverage of action (Biermann et al. 2010: 30), suggesting transnational climate change governance actors are able to shape the context in order to engage a wider variety of non-state actors than the UNFCCC is able to engage. This is indicative of the capacity of transnational climate change actors to shape the context at will, and thereby include a diverse range of responsible actors.

All of the above has allowed for a wide variety of actors to participate in transnational climate change governance, including those that are not interested in the multilateral process (Bulkeley and Newell: 2010: 97). Their participation has meant policy innovation, experimentation and demonstration of effects and best practices, all of which can diffuse transnationally (Sander et al. 2015: 467). These actors have also helped build capacity, established norms of ambitious climate action and catalysed supportive political coalitions, facilitating international cooperation (Sander et al. 2015: 467). In this sense, transnational climate change governance actors are having a profound effect on how climate change is being responded to at the non-state level. All of the above strongly suggests that transnational governance actors have the capacity to create a context within which individuals, firms, sub-state entities and international institutions can act on their responsibilities under Demand Three. Indeed, it seems that this context is well on the way to being created. Nevertheless, as was explained above, actors are typically not held to account more than voluntarily within transnational climate change governance. Of course, voluntariness does not suggest that implementing actors will not comply with the rules of initiatives or will fail to act out the initiatives' core functions (Hoffman 2011: 38). Even if an initiative is voluntary, it can still exert a degree of power over its members, in the form of 'soft' regulation, as discussed above (Bulkeley and Newell 2010: 57). However, it is worth re-stressing that transnational climate change governance actors do not usually hold individuals, firms, sub-state entities and international institutions to account more than voluntarily.

To further illustrate that transnational climate change governance actors have the capacity to create a context within which individuals, firms, sub-state entities and international institutions can act on their responsibility for lowering emissions and/or contributing financially to the climate change cause, the chapter will now turn to two illustrative examples: the Verified Carbon Standard and the Regional Greenhouse Gas Initiative. Each of these public initiatives creates a context where non-state actors act on their responsibilities, with the Verified Carbon Standard focusing corporations and the Regional Greenhouse Gas Initiative encompassing sub-state authorities. The chapter has previously discussed that corporations play an important role in climate change action through finance and innovation, and that climate change mitigation and adaptation at the sub-state level (for example in cities) can have a positive impact in terms of setting best practice, knowledge sharing and including the voices of those most affected by climate change. For this reason, both the Verified Carbon Standard and the Regional Greenhouse Gas Initiative present important and useful examples of transnational climate change governance.

Let us first turn to the Verified Carbon Standard (VCS), which was founded by a collection of business and environmental leaders who saw a need for greater quality assurance in voluntary carbon trading markets (VCS 2017a). Carbon trading is a market-based approach to combating climate change that centres around financial incentives for lowering global carbon dioxide (CO_2) emissions levels. It is currently the most common type of emissions trading, which can include the trading of any one or more GHGs (such as methane, chlorofluorocarbons and nitrous oxide). Most carbon trading regimes operate as a 'cap and trade' system. Under this kind of regime, an upper limit of emissions is set, and emissions are then traded within this cap. A country, city or corporation that has high emissions needs can pay for the right to emit more than their allocated share by buying permits from actors with low emissions. This offers a financial incentive for reducing emissions, since selling off emissions permits is profitable. Carbon markets are said to allow emissions cuts at the lowest possible cost, because they do not require adoption of green technology from all actors.

To establish some degree of structure and stability in the world of carbon trading, a number of standards, included the VCS, were established to provide rules that assure buyers that their purchased credits are genuinely reducing GHG emissions (Stevenson and Dryzek 2014: 110). The VCS has captured the largest share of this market: it is the world's most widely used voluntary GHG reduction program (VCS 2017c). The VCS does not buy and sell carbon credits, but rather facilitates this exchange by eliminating the need for the purchaser to evaluate the merits of many different projects (Stevenson and Dryzek 2014: 111). In other words, the VCS provides a common standard in the carbon trading market. The role of the VCS is therefore to facilitate carbon trading, rather than to hold corporations to account. Nevertheless, the VCS, by providing a trusted standard of carbon trading, enables corporations to act on the responsibilities outlined in Demand Three, by allowing these actors to buy and sell emissions and display their emission-lowering efforts. The VCS explicitly aims to result in 'massive emission reductions across the world' (VCS 2017b). This indicates that the VCS seeks to assist corporations in lowering their emissions, which enables Demand Three because the VCS creates a context within which corporations are able to act on their responsibilities to lower emissions. According to the VCS, over 300 certified projects have collectively reduced or removed more than 200 million tonnes of CO_2 emissions from the atmosphere, indicating that corporations are acting on their responsibilities on a sizable scale (VCS 2017c).

The Regional Greenhouse Gas Initiative (RGGI) is another good example, which illustrates that transnational governance actors have created a context where non-state actors can comply with their climate-related responsibilities. The RGGI is comprised of sub-state actors in the north-east and mid-Atlantic regions of the USA: Connecticut, Delaware, Maine, Maryland, Massachusetts, New Hampshire, New York, Rhode Island and Vermont. These sub-state actors aim to cap and reduce GHG emissions from the power sector (RGGI 2017a). In order to achieve this, RGGI members sell their allowances through auctions and invest proceeds in energy efficiency, renewable energy and other consumer benefit programs (RGGI 2017a). In 2015, proceeds were invested in programmes including energy efficiency, clean and renewable energy, greenhouse gas abatement and direct bill assistance (RGGI 2017b). The RGGI estimates that through these investments, 9 million MWh of electricity use, 28 million MMbtu of fossil fuel use and 5.3 million short tons of CO_2 emissions were avoided.

The RGGI is therefore quite clearly creating a context within which sub-state entities can act on their climate change responsibilities by reducing GHG emissions and investing in climate-friendly technology. The scale of this context is massive: in 2015, over 161,000 households and 6,000 businesses participated in programmes funded by RGGI investments, while 1.5 million households and 37,000 businesses received direct bill assistance (RGGI 2017b). The RGGI is clearly allowing sub-state actors to make a substantial contribution to emissions reductions. This is important, considering that the USA is currently not held to account for emissions reductions or financial contributions under the Kyoto Protocol, and has plans to withdraw from the Paris Agreement. Although the RGGI is a voluntary initiative, it is illustrative of the fact that transnational climate change governance creates a context within which sub-state actors can act on their responsibilities under Demand Three, thereby enabling this demand.

It should be noted that the six examples outlined under the heading of Demands One and Two also illustrate the fact that transnational climate change governance actors are creating a context within which capable actors, including individuals, firms, sub-state entities and international institutions, can act on their responsibilities for lowering emissions and/or contributing financially to the climate change cause. The C40 involves cities attempting to lower emissions, the Climate Group is comprised of cities and corporations attempting to spread clean technology, the Asian Cities Climate Change Resilience Network is a group of cities aiming to strengthen the adaptive capacity, 100 Resilient Cities has corporate and sub-state members who aim to help cities around the world become more resilient to climate change, the Carbon Sequestration Leadership Forum is a group of national governments attempting to lower GHGs by capturing carbon and the Asia Pacific Partnership on Clean Development and Climate was comprised of states who attempted to facilitate technological innovation.

From these examples, and the discussion above, it is clear that transnational climate change governance actors are capable of enabling Demand Three by creating a context within which non-state actors can act on their responsibilities. It is worth reiterating that actors are not held to account in a legally binding manner, but rather voluntarily. Of course, soft compliance measures are not enough to hold these actors fully to account. Nevertheless, transnational climate change governance actors are shaping how individuals, communities, cities, countries, provinces, regions, corporations and states respond to climate change

(Hoffman 2011: 8). The examples discussed throughout the chapter have served to illustrate this. Therefore, actors in transnational climate change governance can be said to reside on the third rung of the hierarchy in terms of enabling Demand Three:

> 3. Actors in the institution have promised to begin working on enabling the demand of justice in the future – no policy has been adopted, but there is the potential for the creation of policy in order to consistently work towards enabling the demand of justice.

Although it cannot be said that all actors within transnational climate change governance have promised to begin working on enabling Demand Three in the future, because this is impossible to claim due to vast number of projects in existence, the discussion above has served to illustrate the potential for policy which enables Demand Three. The potential for policy lies in the fact that there exists a context under which the responsibility of individuals, firms, sub-state entities, international institutions and states to act on climate change could be inscribed into policy. The Verified Carbon Standard, which allows corporations to act on their responsibilities, and the Regional Greenhouse Gas Initiative, which allows sub-state actors to act on their responsibilities, are indicative of this context. In addition, the C40, the Climate Group, Asian Cities Climate Change Resilience Network, 100 Resilient Cities, the Carbon Sequestration Leadership Forum and the Asia Pacific Partnership on Clean Development and Climate all allow various state and non-state actors to act on their climate responsibilities outlined in Demand Three, and are also indicative of this context. Nevertheless, it cannot be said that transnational climate change governance responses have policies in place that consistently enable Demand Three, because actors are not held to account more than voluntarily. For this reason, transnational climate change governance actors remain on the third rung of the hierarchy in terms of Demand Three.

On a final note, it is worth stressing that participation of non-state actors in transnational climate change governance should not necessarily be taken as a sign that these actors consider themselves responsible for acting on climate change. In fact, there is concern within transnational climate change governance research that the pervasiveness of market mechanisms and the lack of innovation beyond 'business as usual' calls into question whether actors are being held to account for their responsibilities or rather pursuing their own interests and agendas.

The bulk of transnational responses to climate change are market-oriented, and the centre of gravity of the global response is bound up with market mechanisms, including credit and allowance markets (Hoffman 2011: 149). Unfortunately, this reinforces the notion that the transition to a carbon-neutral world should and will take place through market-oriented means rather than through radical rethinking of social and economic structures (Hoffman 2011: 40). Such rethinking is necessary to ensure a just response to climate change, because it is not clear how effective the incremental changes that arise from market mechanisms can be. Policies that are relatively easy to implement (such as carbon markets) may be exactly the ones unlikely to make much difference (Stevenson and Dryzek 2014: 4).

Given current scientific knowledge, it is widely agreed that the only hope for permanently avoiding dangerous climate change lies in a rapid transition to a 'low-carbon economy' (Aldred 2016: 150). There is widespread agreement that the transition to such an economy cannot be achieved by emissions trading or environmental taxes alone: policies to encourage technological innovation are also essential (Aldred 2016: 152). By prioritising abatement among producers with the lowest marginal abatement costs, emissions trading encourages adoption of the best available existing technology, rather than innovation to develop new ones (Aldred 2016: 153). Perhaps more troublingly, market mechanisms have been criticised for exacerbating existing inequalities. Research demonstrates that it is the rich who buy credits from the poor, not vice versa, entrenching existing inequalities (Hayward 2007: 439). In addition, market mechanisms (like cap-and-trade systems) are likely to hit poorer households harder than richer households, with unwelcome implications for distributional justice (Caney and Hepburn 2011: 233). This further adds to the idea that market mechanisms replicate 'business as usual' rather than reflecting the genuine structural change that is necessary to combat climate change and ensure a just response to the climate change problem.

For this reason, some researchers suggest that transnational climate change governance initiatives simply help reproduce the norms and rationalities and further the interests of dominant political economic forces such as global finance (Bulkeley et al. 2015: 59). This is a compelling argument, especially considering that there were no broad-based anti-consumption movements or anti-consumerism movements and campaigns within Bulkeley et al.'s sample of sixty projects (Bulkeley et al. 2015: 101). Also absent were campaigns that demonise and seek to eliminate coal use, punish those who invest in coal-burning facilities or reward those that significantly reduce beef or meat consumption (Bulkeley et al. 2015: 101). Overall, it is clear from Bulkeley et al.'s findings that many of the actors who wield financial and political power in transnational climate change governance initiatives have very little interest in fundamentally changing the global response to climate change. This is problematic, because a condition of climate justice cannot be reached under the scope of current climate change governance action, both multilateral and transnational.

However, it should be noted that there has been an emergence of critical transnational climate change governance initiatives that represent a shift from 'business as usual' and provide further scope for optimism. These initiatives adopt a strong focus on green ideology, social justice and the moral imperative of climate change. Two examples of this type of initiative are Carbon Trade Watch and Transition Towns. Carbon Trade Watch emerged as a critical non-governmental organisation attempting to undermine the legitimacy of carbon markets as a credible response to climate change. The initiative works to expose the various political and environmental problems produced by such markets, both in the intergovernmental and transnational forms. Interestingly, one of the issue areas Carbon Trade Watch concerns itself with is climate justice (Carbon Trade Watch 2017). This suggests that there are initiatives that are actively seeking to enable a just response to climate change by creating a context where this is possible.

Transition Towns are another example of a counter-status-quo initiative. The Transition Town movement 'explores and develops ways we can change from energy-hungry ways of living that are utterly dependent on oil and other fossil fuels, to ways of living that are significantly less so' (Transition Towns 2017). The Transition Town Movement calls for a fundamental shift in patters of transportation, energy use, food production and personal lifestyles, attempting to address some of the root causes of climate change rather than tinker at the margins. This is a critical approach, which does not simply perpetuate the status quo. The two examples above serve to illustrate that transnational climate change governance is not as one-sided as critics suggest. Nevertheless, actors in transnational climate change governance will need to address the problem of 'business as usual' if climate change action is to involve more than a very slight change in behaviour which perpetuates existing inequalities. This will be further discussed in the concluding section of the chapter.

CONCLUSION: SUMMARY OF FINDINGS

The chapter has now completed its assessment of the transnational climate change response. What is perhaps most striking about this assessment is that none of the three demands of climate justice have been fully enabled. In fact, transnational actors sit on the third rung of the four-point hierarchy in the case of each demand:

The Four-Point Hierarchy

1. Actors in the institution enable the demand of justice – the demand of justice is unequivocally fulfilled in its entirety.
2. Actors in the institution are consistently working towards enabling the demand of justice – the demand of justice is not yet fulfilled, but there are policies in place which are consistently leading towards this goal.
3. Actors in the institution have promised to begin working on enabling the demand of justice in the future – no policy has been adopted, but there is the potential for the creation of policy in order to consistently work towards enabling the demand of justice.
4. Actors in the institution do not enable the demand of justice – there has been no promise or attempt to enable the demand of justice and there are no policies in place.

Although these results are not indicative of a just transnational climate change regime, it is nevertheless encouraging to see that transnational actors have moved beyond the bottom rung of the hierarchy. It is worth pausing to appreciate that these actors are working towards enabling all three demands of climate justice:

The Three Demands of Climate Justice

1. a) Global temperature changes must be kept at or below 2°C.
   b) Adaptation must be prioritised alongside mitigation.
2. The distribution of benefits and burdens in global climate change action must be based on the Polluter's Ability to Pay (PATP) model.
3. Capable actors, including individuals, firms, sub-state entities and international institutions, irrespective of the country in which they live or exist, must be held responsible for lowering emissions and/or contributing to adaptation efforts, in line with their respective capabilities.

In terms of the first part of Demand One, the assessment above suggests that the nature of transnational climate change governance processes allows for increased participation, makes it possible to target emissions outside of the UNFCCC and raises ambitions for lowering emissions. Some initiatives, such as the C40 and Climate Group, explicitly aim to lower emissions, all of which creates a context where the first part of Demand One can be enabled. In terms of the second part of Demand One, although only very few initiatives, including the Asian Cities Climate Change Resilience Network and 100 Resilient Cities, explicitly focus on adaptation, there is some room for optimism that this will change over time, considering the most recent studies only assess initiatives the existed before 2013 and the fact that existing methodologies may have excluded some initiatives. In addition, there have been promises made, specifically through the 2014 Global Investor Statement on Climate Change, to raise money for adaptation and enable technology transfer and assistance.

In terms of Demand Two, although states are not the primary actors in transnational climate change governance, those that do participate go beyond information sharing or capacity building and engage in certification or target setting, which indicates that states are being held to account, even if this is through soft, voluntary measures (Bulkeley et al. 2015: 35). Furthermore, the assessment illustrated that some existing initiatives enable the PATP model. This includes the Carbon Sequestration Leadership Forum, which specifically targets high-emitting and/or wealthy countries that can contribute to funding new technologies, and the Asia Pacific Partnership on Clean Development and Climate, where members contributed financially to lower global emissions through technology change.

Finally, in terms of Demand Three, the assessment above revealed that transnational climate change governance actors are shaping how individuals, communities, cities, countries, provinces, regions, corporations and nation-states respond to climate change (Hoffman 2011: 8). These responses include the Verified Carbon Standard, where corporations act on their responsibilities by trading carbon, and the Regional Greenhouse Gas Initiative, where sub-state actors act on their responsibilities by reducing GHG emissions and investing in climate-friendly technology.

These findings indicate that transnational climate change actors are moving towards enabling a condition of climate justice. It is therefore important that climate justice scholars continue to research transnational climate change governance processes. As Bulkeley et al. (2015: 3) put it, 'all individuals concerned about climate change – academics, activists, citizens and policymakers – should be interested in how transnational climate change governance works and how it might contribute to efforts to stave off the consequences of climate change'. This is especially true when considering the potential these actors have. For one, there is no indication that specific types of actors (non-governmental organisations, corporations, states and so on) engage with particular interest or issue areas. In this sense, transnational climate change governance actors have created a flexible context within which actors can pursue a myriad of interests (Keohane and Victor 2011: 20). Second, transnational climate change governance provides room for innovation. Transnational actors push the envelope of what is possible, actively seeking out gaps in the response to climate change and attempting to fill them (Hoffman 2011: 78). Continued innovation has the potential to catalyse change, which may provide the momentum necessary to adequately respond to the climate change problem (Ostrom 2009: 4). Third, transnational climate change governance actors may be able to provide a source of friction that catalyses a demand for a broad transformation in the response to climate change (Hoffman 2011: 28). Through this, transnational climate change governance processes may be able to create significant momentum for the global response. Finally, recent research indicates that transnational climate change governance will continue to grow and expand (Bulkeley and Newell 2011: 110). This momentum for growth and expansion is important because it creates even more room for the three demands of justice to be pursued at the transnational level. For all of these reasons, transnational climate change governance should be taken as a source of optimism at a time when the prospects for a just global response to climate change are at best uncertain.

Nevertheless, it should be stressed that there remain problems in the case of each demand of justice. In terms of Demand One, transnational climate change governance actors have not delivered anything remotely close to the emissions reductions required to protect the human right to health. This has led some critics to claim that transnational climate change governance actors are merely pursuing their own interests, rather than making fundamental changes to the global response to climate change. It is difficult to discern whether this is the case because transnational initiatives often do not set emissions targets, and, if they do, emission reductions are challenging to measure and there is very little evidence on whether existing measurements are reliable. The second part of Demand One also faces problems. Transnational actors appear to strongly favour mitigation over adaptation, much like multilateral actors. And, when considering specific adaptation measures such as financing and technology transfer, a similar picture emerges. There is currently clearly not enough focus on adaptation, and even when there is, the most vulnerable are not being prioritised (Sander et al. 2015: 467). Transnational climate change actors must become much more focused on adaptation if the second part of Demand One has any hope of being realised.

In terms of Demand Two, it is difficult to claim with any certainty that transnational climate change governance actors create a context within which states are held to account in line with the PATP model. This is for two reasons. The first is that states are not the primary actors within transnational climate change governance, and the second is that even when states are the primary actors within a project, they are not often held to account more than voluntarily. In addition, transnational climate change governance seems to favour actors from a certain area of the world: namely Western, developed nations, primarily the USA, Canada and Australia. Less developed countries present a minority, and the most vulnerable regions of Sub-Saharan Africa (SSA), Oceania, and the Middle East and North Africa (MENA) remain particularly under-represented (Bulkeley et al. 2015: 123). In this sense, a version of the traditional North–South divide found in multilateral response is being replicated at the transnational level, with more participation and leadership from developed than from developing countries (Hale 2016: 20).

Finally, in terms of Demand Three, there is concern within transnational climate change governance research that the pervasiveness of market mechanisms and the lack of innovation beyond 'business as usual' calls into question whether actors are being held to account for their responsibilities, or rather pursuing their own interests and agendas. The fact that transnational climate change actors fail to fully enable any of the three demands represents a key hindrance that must be overcome to meet a condition of justice in the case of climate change. Transnational actors must do more to live up to their responsibility for enabling a condition of justice.

Interestingly, many of the problems discussed in this chapter overlap with the problems faced by multilateral actors. In terms of Demand One, both multilateral and transnational actors face the problem of holding enough actors to account for emissions reductions. In addition, both types of climate change response place much more emphasis on mitigation than adaptation. In terms of Demand Two, both multilateral and transnational actors are struggling to enable a PATP model of responsibility and are failing to include less developed countries in decision-making processes in a fair and representative manner. Finally, in terms of Demand Three, although transnational actors are seemingly involving many non-state actors, and multilateral actors are moving in the direction of including non-state actors under the UNFCCC's regime, both sides of the climate change response are struggling to change the behaviour of these actors beyond 'business as usual'.

These common problems suggest that there is some continuity in global climate change governance, even though multilateral and transnational climate change responses are quite different in their scope and ambition. In this sense, the cosmopolitan assessment of the climate change response revealed that it is not possible to claim that multilateral governance has failed, and transnational governance presents improvement and innovation: instead, these processes face similar problems. These shared problems imply that including transnational climate change actors in multilateral processes, which the post-Paris Agreement regime is moving towards, may not be a simple or straightforward improvement of the climate change response. The Conclusion of the book will therefore reflect on whether the direction in which the post-Paris Agreement regime is heading might be a hindrance to a just response to the climate change problem, rather than a help.

NOTE

1. For a Discussion of the PATP, see Chapter 3.

CONCLUSION

This book set out to make sense of the lack of proper response to climate change – focusing on what has gone wrong, what has gone right and what might change now that the Paris Agreement has been ratified. In order to achieve these aims, the book conducted a cosmopolitan assessment of the multilateral and transnational climate change response and in doing so endeavoured to make sense of the 'big picture' of climate change (mis)management and the injustices that come along with it. This endeavour has allowed the book to demonstrate that the theory of climate justice can be applied to evaluate the practice of climate change governance. In other words, the book has demonstrated that the gap between theory and practice can be bridged. Although there is more work to be done, the bridging of climate justice and climate governance has allowed for a comprehensive normative understanding of the climate change response and provided new insights into the governance of climate change. These insights can underwrite future research and ultimately help to bring about a more just global response to climate change, because they have provided a common denominator from which to begin suggesting reform. The Conclusion of this book therefore focuses on the lessons that can be learned from the bridging of theory and practice.

More specifically, the Conclusion will focus on how transnational and multilateral responses compare, and what policy recommendations can be made on the basis of this comparison. As was briefly explained in the previous chapter, multilateral and transnational actors face very similar problems, including the ongoing struggle to lower greenhouse gas (GHG) emissions at the rate required, the entrenched favouring of mitigation over adaptation, the pervasive exclusion of less developed countries from decision-making processes and the incessant failure to change the behaviour of responsible actors. These shared problems imply that including transnational climate change actors in multilateral processes, which the post-Paris Agreement regime is moving towards, may not be a simple or straightforward improvement of the climate change response. The Conclusion will therefore consider whether the direction the post-Paris regime is heading in might be a hindrance to a just response to the climate change problem, rather than a help. First, the Conclusion will provide a brief summary of what the cosmopolitan assessment of multilateral and transnational climate change governance has revealed about the lack of proper response to climate change.

WHAT HAS BRIDGING THEORY AND PRACTICE REVEALED?

The book was divided into two distinct halves, one focusing on theory (climate justice), and the other on practice (climate governance). Part I, Developing a Climate Justice Account, focused on answering four normative questions: who will be most affected, what exactly is at stake, what action must be taken in the face of climate change and who should be responsible for this action. In answering these questions, the book set out the scope, grounds and demands of climate justice in Chapters 1, 2 and 3, respectively. Part II, Assessing Climate Governance, then focused on how multilateral and transnational climate change actors have fared in meeting the demands of climate justice set out in Part I. Chapter 4 explained how the global response to climate change was to be assessed, and Chapters 5 and 6 then turned to the assessment of multilateral and transnational climate change responses, respectively. All six chapters are briefly summarised in more detail below. This summary serves to remind readers of what has been found before the Conclusion makes normative recommendations on the basis of these findings.

Chapter 1 focused on who will be most affected, and put forward that relational and non-relational elements are necessary to fully include the primary victims of climate change in the scope of climate justice. Determining what future generations are owed requires non-relational analysis, whereas exploring what is owed to less developed countries requires relational analysis. Chapter 1 therefore made the case for an account that is both relational and non-relational. This mixed account is based, in the first instance, on a non-relational scope that encompasses every individual, including future individuals. The relational element of the account can then be applied in order to explore existing climate change relationships – for example, the relationship between developed and less developed countries. In this sense, the mixed account uses the non-relational side of the account as a foundation that defines a global scope and establishes a minimal moral threshold, and uses the relational side of the account to define demands of justice based on the relationships created by climate change.

The mixed account was further developed in Chapter 2, which focused on what is at stake by defining the non-relational grounds of climate justice: the human right to health. Chapter 2 first set out reasons for using a human rights approach, explaining that such an approach is commonly used to define moral thresholds, is recognised as important by policymakers at the global level, 'humanises' the climate change problem, adds urgency to the debate and is an increasingly popular approach among climate justice scholars. The chapter then defended the prioritisation of health over other human rights by illustrating that it captures the threats climate change poses to humans more comprehensively than other key human rights, including the right to food and water, the right to life and the right to free movement. Next, Chapter 2 provided a definition of the right to health: the right to a standard of living adequate for health, including the right to be adequately nourished and have adequate shelter. Finally, the chapter began to explain what action is necessary in order to protect this right.

Chapter 3 expanded on this discussion and focused on what action must be taken in the face of climate change, and who should be responsible for this action. In answering these questions, Chapter 3 specified what exactly is normatively expected of the climate change response by developing three demands of climate justice. Demand One centred around a minimum set of actions that must be taken to protect the non-relational human right to health. It was argued that protecting this right will require policymakers to, at the very minimum, keep global temperature changes to 2°C or less. It was also explained that adaptation is key for protecting human health, because strengthening health systems could significantly reduce the burden of disease and promote population health. The second and third demands of climate justice centred on who is responsible for ensuring temperatures are kept to this minimum, and/or who is responsible for contributing to adaptation efforts. To determine the responsibilities of states, the chapter conducted relational analysis of the relationship between developed and less developed countries, exploring what a fair distribution looks like in the context of this relationship. Ultimately, Chapter 3 recommended the Polluter's Ability to Pay approach, or PATP, as a model for climate responsibility. Finally, the chapter determined the responsibilities of non-state actors by focusing on the actors causing climate change and exploring how they relate to those who are suffering from climate change effects. Using the relational side of the climate justice account, the chapter put forward that non-state actors must be held responsible for lowering emissions and/or contributing to adaptation efforts, because they contribute to and perpetuate a wider system of injustice which threatens the human right to health of present and future peoples.

The three demands of justice developed in Chapter 3 can be found below. These three demands are considered normative principles that must underwrite a more just global response to climate change.

The Three Demands of Climate Justice

1. a) Global temperature changes must be kept at or below 2°C.
   b) Adaptation must be prioritised alongside mitigation.
2. The distribution of benefits and burdens in global climate change action must be based on the Polluter's Ability to Pay (PATP) model.
3. Capable actors, including individuals, firms, sub-state entities and international institutions, irrespective of the country in which they live or exist, must be held responsible for lowering emissions and/ or contributing to adaptation efforts, in line with their respective capabilities.

Once the above demands were set out, the book turned to assessing to what extent they have been enabled by the global response to climate change. Chapter 4 served as a conceptual introduction to the assessment of climate change governance. After discussing the importance of evaluating both multilateral and transnational climate change governance actors, Chapter 4 put forward that these actors have a moral responsibility to enable a condition of climate justice, due to their capability of creating a social and political context within which the three demands of justice defended in this book can be met. Moreover, Chapter 4 argued that multilateral actors under the United Nations Framework Convention on Climate Change (UNFCCC) have formal authority to act, and are therefore more responsible for enabling a condition of justice in the case of climate change. However, this does not diminish the moral responsibility of other actors, specifically those involved in transnational governance processes, should the actors under the UNFCCC should fail to enable the three demands of justice. Finally, Chapter 4 explained how the book would conduct its assessment of climate change governance by setting out four-point hierarchy. This hierarchy can be found below.

The Four-Point Hierarchy

1. Actors in the institution enable the demand of justice – the demand of justice is unequivocally fulfilled in its entirety.
2. Actors in the institution are consistently working towards enabling the demand of justice – the demand of justice is not yet fulfilled, but there are policies in place which are consistently leading towards this goal.
3. Actors in the institution have promised to begin working on enabling the demand of justice in the future – no policy has been adopted, but there is the potential for the creation of policy in order to consistently work towards enabling the demand of justice.
4. Actors in the institution do not enable the demand of justice – there has been no promise or attempt to enable the demand of justice and there are no policies in place.

After the parameters of assessment were clarified, the book turned to assessing to what extent multilateral actors enable the three demands of justice in Chapter 5. In doing so, Chapter 5 aimed to provide a broad-brush overview of multilateral climate change governance. Taking each demand of justice in turn, the chapter focused on the normative commitments made in the Convention, assessed the policies set out in the Kyoto Protocol, examined what has been achieved so far by multilateral actors and considered to what extent the Paris Agreement presents a change from existing policies. In this way, Chapter 5 provided a historical overview of multilateral climate change action, as well as looking to the future. Encouragingly, the assessment conducted in Chapter 5 revealed that there is movement forward in the case of each demand of justice.

In terms of the first part of Demand One, the Paris Agreement has resulted in more countries than ever before pledging to lower emissions, which means that the right to health has a higher chance of being protected than under the Kyoto Protocol. In terms of the second part of Demand One, the multilateral regime has made progress by monitoring adaptation needs through National Adaptation Programmes for Action (NAPAs), National Adaptation Plans (NAPs) and Intended Nationally Determined Contributions (INDCs), and the Paris Agreement appears to move forward on adaptation policy, at least in terms of what has been promised. The creation of the Least Developed Experts Group (LEG) and Adaptation Committee are also encouraging, because they create space for adaptation to be prioritised alongside mitigation.

In terms of Demand Two, although the Paris Agreement was not able to move forward on the outdated categorisation of Annex I and II, the INDCs have, for the first time, ensured near-universal participation in the climate change regime. There are now more rich and high-emitting countries being held to account than ever before, albeit under weak enforcement mechanisms. The focus on INDCs and NAPAs/NAPs rather than a top-down regime also indicates a shift in how less developed countries have been included in decision-making. Although their preferred interpretation of Common but Differentiated Responsibility (CBDR) was ultimately not included in the Paris Agreement, these countries have more of a say in the multilateral regime than ever before.

Finally, in terms of Demand Three, the run-up to the Paris Agreement and the negotiations at COP21 represent a significant shift in the perception of the role of non-state climate change actors. Although the Paris Agreement itself does not make mention of these actors, it was clear from the negotiations at COP21 that non-state actors are expected to play an increasingly important role in the post-Paris Agreement regime. In fact, many of the INDCs include mention of non-state actors, and it is clear that multilateral actors are beginning to shift their attitude towards these actors.

Despite these positive findings, Chapter 5 also revealed that there is still an urgent need for change in the multilateral regime. None of the three demands were enabled past the third rung on the four-point hierarchy. This is due to a number of key problems. For one, the targets set out in the Kyoto Protocol are unequivocally not in line with what is required to protect the human right to health. The small number of countries held to account until 2020 only account for 15% of global emissions, and their efforts will not make enough impact to protect future generations. In addition, although the Paris Agreement has increased mitigation ambitions by setting a 1.5°C temperature change goal, the current INDCs are not in line with this target. In fact, global temperatures are set to warm by between 2.7°C and 4.9°C, reaching 3.6°C by 2100 under plans set out in the current INDCs. Furthermore, although the Paris Agreement has specified new compliance mechanisms, these remain voluntary and non-punitive, calling into question to what extent states will be held to account for their INDCs.

The story is largely the same for adaptation. Although the creation of an Adaptation Committee, LEG and the focus on NAPAs are all signs of progress, the funding mechanisms set out in the Kyoto Protocol have so far failed to raise adequate finance for adaptation, and less developed countries remain frustrated at the slow pace of technology transfer. In addition, NAPAs have not yet been acted on, and the Paris Agreement did not set out new financial targets. Finally, financial contributions and technology transfer remain voluntary, calling into question whether the pace of adaptation will ever be adequate for the protection of the human right to health.

The Kyoto Protocol also unequivocally does not enable the PATP model of responsibility, because it fails to include many of the richest and highest-emitting countries. And, although the Paris Agreement has moved forward on this problem with the creation of INDCs, there are still disagreements over who should be responsible for what, with developed countries favouring an Ability to Pay (ATP) model and less developed countries preferring a Polluter Pays (PPP) model. It is unclear which model will be pursued, and to what extent rich and high-emitting countries will be expected to contribute more than poorer and low-emitting countries. Furthermore, although less developed countries are increasingly able to report on their needs, their voices are still often left out of climate change negotiations. There is hope for improvement on this in future, especially considering that the Least Developed Country Work Programme aims to provide training in negotiating skills and language and to develop the capacity of negotiators from Least Developed Countries to participate effectively in the climate change process. However, to what extent they will be included in negotiations in the run-up to the Paris Agreement being implemented remains to be seen.

Finally, although the negotiations at COP21 included non-state actors more seriously than ever before, the Paris Agreement ultimately made no mention of these actors. This leaves open the question of what is next, or more exactly the question of whether multilateral actors will create a context where non-state actors can be held to account for their climate related responsibilities. Overall, then, multilateral actors have certainly made progress, but also have a long way to go to ensure a just response to the climate change problem.

After concluding the assessment of multilateral climate change governance in Chapter 5, the book moved on to assessing to what extent transnational actors enable the three demands of justice in Chapter 6. In doing so, Chapter 6 aimed to provide a broad-brush overview of transnational climate change governance, to allow readers to develop an understanding of the vast and rapidly changing non-state climate change response. Chapter 6 made use of both existing climate change governance research and ten examples of transnational climate change governance initiatives: the C40, the Climate Group, the Asian Cities Climate Change Resilience Network, 100 Resilient Cities, Carbon Sequestration Leadership Forum, the Asia Pacific Partnership on Clean Development and Climate, the Verified Carbon Standard, the Regional Greenhouse Gas Initiative, Carbon Trade Watch and Transition Towns. Chapter 6 focused on one demand of climate justice at a time, assessing both what has been promised by transnational actors and what has been achieved so far. The assessment revealed that there is room for cautious optimism because there is movement forward in the case of each demand of justice.

In terms of the first part of Demand One, the nature of transnational climate change governance processes allows for increased participation, makes it possible to target emissions outside of the UNFCCC and raises ambitions for lowering emissions. Some initiatives, such as the C40 and Climate Group, explicitly aim to lower emissions, suggesting there is a context being created where emissions can be kept in check. In terms of the second part of Demand One, Chapter 6 found that although only very few initiatives, including the Asian Cities Climate Change Resilience Network (ACCRN) and 100 Resilient Cities (100RC), explicitly focus on adaptation, there is some room for optimism that this will change over time, considering the age of the most recent studies and the fact that existing methodologies may have excluded some initiatives that have come into being in the last few years. In addition, there have been promises made, specifically through the 2014 Global Investor Statement on Climate Change, to raise money for adaptation and enable technology transfer and assistance.

In terms of Demand Two, although states are not the primary actors in transnational climate change governance, those that do participate go beyond information sharing or capacity building and engage in certification or target setting, which indicates that they are being held to account, even if this is through soft, voluntary measures. Furthermore, the assessment in Chapter 6 illustrated that some exiting initiatives enable the PATP model. These include the Carbon Sequestration Leadership Forum (CSLF), which specifically targets high-emitting and/or wealthy countries that can contribute to funding new technologies, and the Asia Pacific Partnership on Clean Development and Climate (APP), where rich and high-emitting members contributed financially to lower global emissions through technology change.

Finally, in terms of Demand Three, Chapter 6 revealed that transnational climate change governance actors are shaping how individuals, communities, cities, countries, provinces, regions, corporations and nation-states respond to climate change. The Verified Carbon Standard (VCS), which allows corporations to act on their responsibilities, and the Regional Greenhouse Gas Initiative (RGGI), which allows sub-state actors to act on their responsibilities, are examples of this shaping taking place.

Despite these positive findings, there is still significant room for improvement in the transnational response to climate change. None of the three demands were enabled past the third rung on the four-point hierarchy. This is due to a number of key problems that are not dissimilar to the problems faced by the multilateral regime. For one, transnational actors have not delivered anything remotely close to the emissions reductions required to protect the human right to health. This has led some critics to claim that transnational climate change governance actors are merely pursuing their own interests, rather than making fundamental changes to the global response to climate change. It is difficult to discern whether this is the case because transnational initiatives often either do not set emissions targets or, if they do, the emissions reductions are challenging to measure, and there is very little evidence whether existing measurements are reliable. Furthermore, transnational actors appear to strongly favour mitigation over adaptation, much like multilateral actors. There is currently not enough transnational focus on adaptation, and even when there is, the most vulnerable are not being prioritised. In addition, it is difficult to claim with any certainty, that transnational climate change governance actors create a context within which states are held to account in line with the PATP model. This is for two reasons. The first is that states are not the primary actors within transnational climate change governance, and the second is that even when states are the primary actors within a project, they are not often held to account more than voluntarily.

It also appears that transnational actors exclude those from less developed countries. Transnational climate change governance seems to favour actors from certain areas of the world: namely Western, developed nations, primarily the USA, Canada and Australia, with less developed countries, and the regions of Sub-Saharan Africa (SSA), Oceania and the Middle East and North Africa (MENA) remaining particularly underrepresented. It is very clear that a version of the traditional North–South divide is being replicated in transnational climate change governance, with more participation and more leadership from developed than from developing countries.

Finally, the pervasiveness of market mechanisms and the lack of innovation beyond 'business as usual' clearly demonstrates that many of the actors who wield financial and political power in transnational climate change governance initiatives have little interest in fundamentally changing the global response to climate change. However, there has been an emergence of critical transnational climate change governance initiatives that represent a shift from 'business as usual' and provide scope for optimism. These initiatives, for example Carbon Trade Watch and Transition Towns, adopt a strong focus on green ideology, social justice and the moral imperative of climate change. Nevertheless, actors in transnational climate change governance will need to address the problem of 'business as usual' if climate change action is to involve more than a very slight change in behaviour that perpetuates existing inequalities. Overall, then, there is much more work to be done to respond to climate change in a just manner.

The fact that multilateral and transnational climate change actors both fail to fully enable any of the three demands represents a key hindrance that must be overcome to meet a condition of justice in the case of climate change. Both types of actors must do more to live up to their responsibility for enabling a condition of justice.

Importantly, the bridging of theory and practice has revealed that state and non-state responses face similar problems, including an ongoing struggle to lower GHG emissions at the rate required, the entrenched favouring of mitigation over adaptation, the pervasive exclusion of less developed countries from decision-making processes and the incessant failure to change the behaviour of responsible actors. These shared problems imply that including transnational climate change actors in multilateral processes, which the post-Paris Agreement regime is moving towards, may not be a simple or straightforward improvement of the climate change response. The remainder of the Conclusion will therefore reflect on whether the direction the post-Paris regime is heading in might be a hindrance to a just response to the climate change problem, rather than a help.

WHAT NEXT FOR THE CLIMATE CHANGE RESPONSE?

Bridging climate justice and climate governance has revealed that although multilateral and transnational climate change actors face their own difficulties, they overwhelmingly face overlapping problems. The question to ask, then, is why this is happening. If this question is answered, perhaps their common problems can be overcome, resulting in a more just response to the climate change problem. There is not enough room to properly explore all the possible causes of overlapping climate governance problems in this short Conclusion. However, there is time to explore one potential reason that multilateral and transnational responses face similar problems: they heavily influence one another.

Although transnational processes have not been directly linked to the UNFCCC until very recently, existing research suggests that non-state climate actors have always been influenced by the multilateral climate change response (Weiss 2013: 166). As one example, take the preference for market mechanisms among transnational climate change responses. It is said that this preference stems from the fact that multilateral climate change governance has made market mechanisms the dominant solution to the climate change problem (IPCC 2014b: 1031). This may be why transnational climate change governance actors have not yet succeeded in offering an innovative and radical alternative to the current response to climate change, despite their significant potential to do so. At the same time, existing research suggests that transnational responses to climate change have shaped and influenced the evolution of the UNFCCC (Hoffman 2011: 9). It is therefore no coincidence that the post-Paris Agreement climate change regime regards non-state actors as a core element of the climate change response. This perception has most likely been influenced by transnational actors, who have regarded themselves as a central part of the climate change response for years, during which time they organised side meetings at Conferences of the Parties (COPs) and attempted to lobby multilateral actors to include them in formal proceedings.

If multilateral and transnational climate governance actors currently influence one another, and both sets of actors struggle with overlapping problems under this mutual influence, it may be worthwhile to explore whether or not increasing ties between these processes is to be endorsed. In other words, it is worth asking whether multilateral and transnational climate change responses should become more integrated, or whether it is more useful, in terms of enabling a condition of justice, to leave them as separate, or fragmented, processes. The multilateral regime is currently heading in the direction of integration, as was explained in Chapter 5.

The question to ask, then, is whether the current plan for the post-Paris Agreement regime will allow transnational and multilateral actors to face their shared problems together, or whether further integration will make overcoming these problems more difficult. Perhaps the current level of integrations is mutually reinforcing shared problems, meaning that further integration would exacerbate these problems and ultimately lead to an even more unjust climate change response. To explore this idea further, the Conclusion will now briefly examine arguments for and against integration and determine whether one option is potentially more appealing considering what has been found in the climate-justice-focused assessment of climate governance. This brief discussion does not provide a definitive answer, of course. It is instead intended to spur on the conversation between theory and practice that this book has initiated.

According to pro-integration scholars, an integrated climate change regime could maximise the number of actors responding to climate change, which should speed up the rate of mitigation and ensure that money for adaptation becomes available as quickly as possible (Hof et al. 2010, Bulkeley and Newell 2010). This is an intuitively appealing argument, considering that the climate change problem requires a fast and effective response, which has so far eluded climate governance actors. Furthermore, pro-integration scholars claim that if climate change action is coordinated more centrally than it is now, this may ensure that the thousands of transnational projects in existence avoid overlap as much as possible and provide structure for transnational actors to work together more than they do now. Scholars claim that without such a framework, transnational climate governance runs the risk of failing to deliver meaningful results (Sander et al. 2015: 466). Again, this seems intuitively appealing, because coordination could protect the human right to health more so than an ad-hoc and disorganised climate change response.

The most optimistic proponents of an integrated climate change response suggest that strategically linking the multilateral and transnational spheres could draw on the strengths of both. An integrated regime could, for example, encourage the flexibility, innovation and diversity of non-state actors under the legitimacy and global scope of the UNFCCC. At the same time, an integrated regime could address the weaknesses of each process – the lack of central direction of non-state actions and the slow pace and rigidity of the UNFCCC process (Sander et al. 2015: 468). In this way, further integration could enable multilateral and transnational actors to achieve more together than either could on their own (Betsill et al. 2015: 7). But how exactly would this look? Hale (2016: 18) has suggested that an integrated regime could facilitate centralised benchmarking and tracking to promote implementation and the diffusion of best practices, and limit greenwashing wherever possible. This does not seem entirely implausible. The annual COPs have already established themselves as the main annual 'cross-fertilisation space' for civil society, scientists, businesses and industry from all around the world to rally public attention, network and share best practices (ENB 2015: 45). Perhaps, with some effort over the coming years, an integrated regime that operates around benchmarking and tracking could take form.

Scholars who support further integration certainly make such a regime sound like an appealing option. Their predictions of what an integrated regime would look like leave substantial room for optimism in terms of resolving the problems faced by transnational and multilateral actors, especially the ones they share. For example, an integrated regime (as imagined by the scholars discussed above) could improve accountability for emissions reductions and adaptation finance, with multilateral and transnational actors collaborating on benchmarking and ensuring tracking mechanisms are developed. Furthermore, by working together, multilateral and transnational climate change actors might be able change the behaviour of state and non-state actors because they will have more authority, scope and power to do so. Under such a regime, mitigation and adaptation targets have a greater chance of being met, which would be good news for protecting the human right to health.

However, while integration has been praised by some climate governance scholars, others point out that it is crucial not to overregulate the groundswell of transnational climate change responses, because this might risk diminishing the experimentation and innovation that are among the chief benefits of non-state climate change action (Sander et al. 2015: 470). One merely has to look at the description of the observers at COP23 (ENB 2017: 32), who compared the 'the grey corridors of the intergovernmental area' to the 'bustling atmosphere' of the non-state side events, to recognise that one side is flourishing while the other, operating under entrenched procedures and in an atmosphere thick with the history of sluggish negotiations, struggles to innovate. Given that transnational actors are currently failing to move past 'business as usual', it is very important not to stifle their chance for innovation. It seems unlikely that the multilateral regime would allow non-state actors to pursue radical responses to climate change, especially if this threatens their national interest, or, more widely, the global economic system of growth that is currently benefiting many of the most powerful states acting under the UNFCCC. The multilateral regime has previously been criticised for the inclusion of oil and gas companies under their non-state action framework, which suggests that they may favour less radical approaches (Hale 2016: 16). For this reason, it might be best to leave transnational actors to their own devices, rather than integrating them into the multilateral regime, which could stifle their creativity and prevent more radical approaches from emerging.

Furthermore, some scholars argue that without numerous innovative technological and institutional efforts occurring at multiple scales, it will not be possible to learn which combined sets of actions are the most effective in reducing the long-term threat of climate change (Ostrom 2009: 4). In other words, the more actors working on different aspects of the climate change problem, the more likely it will be to find an effective solution. Transnational climate change actors are especially important in this process of innovation. As was explained in Chapter 6, these actors are often pioneering, pushing the envelope of what is possible and actively seeking out gaps in the response to climate change (Hoffman 2011: 78). If transnational actors were reined in by global rules under an integrated climate governance regime, then this innovation might be stifled, which could slow the process of finding effective solutions to the climate change problem. Transnational climate change actors are aware of this problem. At COP21, some transnational initiatives were dissatisfied with what they saw as excessive centralisation (Hale 2016: 16). Overall, then, it seems that an integrated regime might be detrimental to the innovation and effectiveness of the climate change response.

If readers are not convinced by this argument, consider that we may be better off pursuing a multitude of small-scale solutions rather than waiting for an integrated climate change regime to emerge (Ostrom 2009: 4). Given the decades-long failure of the UNFCCC to reach agreement on the most fundamental issues, including who is responsible for how much climate change action, waiting for multilateral actors to coordinate transnational climate change responses may be counterproductive. One merely has to observe the current state of negotiations in the lead-up to the Paris Agreement's implementation to recognise that its ratification has done little to overcome the disagreements that have slowed down climate change action at the multilateral level for decades.

At COP22 in Marrakesh, for example, Parties struggled to move ahead on fundamental issues regarding the implementation of the Paris Agreement. They were impeded in part by the fact that the Paris Agreement did not include any clear differentiation between less developed and developed countries. As was explained in Chapter 5, Parties had initially hoped to create new categories to replace Annex I and II but failed to do so at COP21. Instead, countries were left to decide on their own contributions in the form of INDCs. This caused problems at COP22, where the main discussion of the Ad-Hoc Working Group on the Paris Agreement (APA) focused on how to communicate INDCs effectively, rather than on who should contribute how much (ENB 2016: 15). Meanwhile, the fact that the current pledges are inadequate to stay below 2°C remained unaddressed (ENB 2016: 37). The discussions of the APA clearly demonstrated that important political misalignments remain, particularly with regard to the differentiation of responsibilities (ENB 2016: 38). In this sense, COP22 confirmed fears that the constructive ambiguity around state responsibility in the Paris Agreement (discussed at length in Chapter 5) will continue to negatively affect the pace of implementation (ENB 2016: 38).

Problems continued at COP23 in Bonn/Fiji. Countries failed to agree on how to provide guidance for INDCs. Less developed countries suggested that differing capacities need to be reflected in INDC guidance (ENB 2017: 13). Developed countries agreed on the need for a differentiated approach but rejected 'bifurcation' of responsibilities (ENB 2017: 13). In the end, no agreement on guidance was possible, which means the issue of responsibility has been put off until COP24 in late 2018, slowing the pace of INDCs being updated in light of their inadequacy to meet the 1.5°C target by the time the Paris Agreement comes into force in 2020. Parties also disagreed on how to communicate adaptation needs in their INDCs, with less developed countries wanting more guidance, and developed countries claiming that the Paris Agreement provided enough guidance (ENB 2017: 13). Less developed countries also expressed concern about the slow rate of progress on the Adaptation Fund (ENB 2017: 15). And, although less developed countries succeeded in elevating the Adaptation Fund's prominence in the post-2020 regime, the governance and funding arrangements of the Fund remain unclear (ENB 2017: 31).

Overall, then, the latest negotiations do not seem to indicate a rapid improvement in multilateral deliberations on climate change. In fact, many of the same issues that have plagued negotiations for decades continue to cause problems, despite the Paris Agreement's ratification. If the multilateral regime cannot overcome fundamental problems surrounding mitigation and adaptation that have been in existence for nearly three decades, how can they be expected to effectively coordinate non-state climate change responses? Even multilateral actors themselves have noted the difficulties of bringing non-state actors into a state-led process, which perhaps explains the lack of enthusiasm demonstrated by Parties at COP23 in engaging with the possibility of further integration (ENB 2017: 32). A successful integrated regime, coordinated by the UNFCCC, therefore seems more than unlikely.

There are two final points in favour of fragmentation to consider. The first is that an integrated regime might allow multilateral actors to sidestep their own obligations, lowering the overall global ambition of mitigation and adaptation goals (Hsu et al. 2015: 501). Many point to the past to support these claims, suggesting that an over-reliance on a coalition of prominent environmental non-governmental organisations and big-industry executives stymied the USA's national cap-and-trade legislation, by diverting attention away from securing political support (Hsu et al. 2015: 501). More globally, the closest precedent in world politics to an integrated environmental regime is the range of 'Type II Partnerships' that were incorporated into the World Summit on Sustainable Development in Johannesburg in 2002, and then again at the Rio+20 summit in 2012. The track record of these partnerships is decidedly mixed, with one study reporting that 38% are completely inactive (Hale 2016: 17). This suggests that the optimistic assumptions about increased ambition, emissions reductions and adaptation funding that pro-integration scholars have made are somewhat naïve.

The final point to be made in favour of fragmentation is that transnational responses are more likely to allow us to overcome the fixation on national interest that has prevented the UNFCCC from pursuing meaningful climate change action for decades. Harris (2013: 202), for example, has noted that when transnational actors meet, they routinely focus their attention on the relationships between humans and climate change, including the human suffering that is predicted to occur. As was explained in Chapter 1, if the object of climate policy is to prevent severe human suffering, it may become more difficult for governments to avoid action (Harris 2013: 135). It may also make it easier to mobilise substantial adaptation funding, increasing the likelihood of setting sound mitigation targets (ICHR 2008: vii). For this reason, it is arguably counterproductive to integrate transnational actors into the multilateral regime because this may mean that the human focus is obscured, and national interests are not overcome. This could slow down the pace of the climate change response. Of course, transnational actors could have a positive effect on multilateral actors, but this is not guaranteed. The UNFCCC has been in existence for nearly three decades, and multilateral actors seem very much set in their ways, judging by the slow pace of change in the multilateral climate change response.

In sum, then, although it would be ideal if state and non-state responses could work together to overcome their shared problems, this seems highly unlikely in the current political context. So, for now, the book recommends that transnational actors are given as much space as possible to pursue their ambitions, with limited guidance from the UNFCCC. This will hopefully allow transnational climate change responses to continue to innovate and push boundaries. Perhaps this will even lead to the emergence of more radical climate-justice-focused initiatives. In the meantime, multilateral actors will hopefully come to a decision on key areas of disagreement, including climate responsibility, adaptation costs, technological and financial transfer, and work to ensure that less developed countries become more included in decision-making procedures. In doing so, they will hopefully influence the transnational regime to do the same.

No matter what happens next, the ability to understand how state and non-state climate change actors connect, relate and overlap is likely to become more important for both climate justice and climate governance scholars over the coming years (Betsill et al. 2015: 9). It is my hope that these scholars can, in part thanks to this book, continue to bridge the gap between theory and practice, and, in doing so, develop a greater normative understanding of the climate change response. This understanding will, with any luck, one day enable a more just global response to the climate change problem.

REFERENCES

100 Resilient Cities (2017a), 'About Us', <http://www.100resilientcities.org/about-us/> (last accessed 9 August 2018).

100 Resilient Cities (2017b), 'Cities', <http://www.100resilientcities.org/cities/> (last accessed 9 August 2018).

100 Resilient Cities (2017c), 'Our Impact', <http://www.100resilientcities.org/our-impact/> (last accessed 9 August 2018).

Abbott, K. W. (2012), 'The Transnational Regime Complex for Climate Change', *Environment and Planning C: Government and Policy* 30(4), pp. 571–90.

Abeysinghe, A., and S. Huq (2016), 'Climate Justice for Least Developed Countries Through Global Decisions', in C. Heyward and D. Roser (eds), *Climate Justice in a Non-Ideal World*, Oxford: Oxford University Press, pp. 189–207.

Aldred, J. (2016), 'Emissions Trading Schemes in a "Non-Ideal" World', in C. Heyward and D. Roser (eds), *Climate Justice in a Non-Ideal World*, Oxford: Oxford University Press, pp. 148–68.

Asian Cities Climate Change Resilience Network (2017a), 'About Us', <http://acccrn.net/about-acccrn> (last accessed 9 August 2018).

Asian Cities Climate Change Resilience Network (2017b), 'Our Purpose and Value', <https://www.acccrn.net/about-acccrn/purpose-and-value> (last accessed 9 August 2018).

Asian Cities Climate Change Resilience Network (2017c), 'Members', <https://www.acccrn.net/members> (last accessed 9 August 2018).

Asian Cities Climate Change Resilience Network (2017d), 'Initiatives Map', <https://www.acccrn.net/map> (last accessed 9 August 2018).

Asia-Pacific Partnership on Clean Development and Climate (2014), 'Charter', <https://2001-2009.state.gov/g/oes/rls/or/2006/59162.htm> (last accessed 9 August 2018).

Avant, D. D., M. Finnemore and S. K. Sell (2010), 'Who Governs the Globe?', in D. D. Avant, M. Finnemore and S. K. Sell (eds), *Who Governs the Globe?*, Cambridge: Cambridge University Press, pp. 1–34.

Ayers, J., M. Alam and S. Huq (2010), 'Global Adaptation Governance Beyond 2012: Developing Country Perspectives', in F. Biermann, P. Pattberg and F. Zelli (eds), *Global Climate Governance Beyond 2012: Architecture, Agency, and Adaptation*, Cambridge: Cambridge University Press, pp. 223–34.

Bäckstrand, K. (2008), 'Accountability of Networked Climate Governance: The Rise of Transnational Climate Partnerships', *Global Environmental Politics* 8(3), pp. 74–102.

Barry, B. (1989), *Theories of Justice*, Berkeley: University of California Press.

Beitz, C., and R. Goodin (2009), 'Basic Rights and Beyond', in C. Beitz and R. Goodin (eds), *Global Basic Rights*, Oxford: Oxford University Press, pp. 1–24.

Bell, S., and A. Hindmoor (2009), *Rethinking Governance: The Centrality of the State in Modern Society*, Cambridge: Cambridge University Press.

Betsill, M., N. K. Dubash, M. Paterson, H. Van Asselt, A. Vihma and H. Winkler (2015), 'Building Productive Links between the UNFCCC and the Broader Global Climate Governance Landscape', *Global Environmental Politics* 15(2), pp. 1–10.

Bevir, M. (2009), *Key Concepts in Governance*, London: Sage.

Biermann, F., and I. Boas (2010), 'Preparing for a Warmer World: Towards a Global Governance System to Protect Climate Refugees', *Global Environmental Politics* 10(1), pp. 60–88.

Biermann, F., P. Pattberg, F. Zelli and H. Van Asselt (2010), 'The Consequences of a Fragmented Climate Governance Architecture: A Policy Appraisal', in F. Biermann, P. Pattberg and F. Zelli (eds), *Global Climate Governance Beyond 2012: Architecture, Agency, and Adaptation*, Cambridge: Cambridge University Press, pp. 25–34.

Bodansky, D., (1993), 'The UN Framework Convention on Climate Change: A Commentary', *Yale Journal of International Law* 18, pp. 451–558.

Bowen, K., and S. Friel (2012), 'Climate Change Adaptation: Where Does Global Health Fit in the Agenda?', *Globalization and Health* 8(10), pp. 1–7.

Broome, J. (2012), *Climate Matters: Ethics in a Warming World*, New York: W. W. Norton and Company.

Bulkeley, H., and P. Newell (2010), *Governing Climate Change*, London: Routledge.

Bulkeley, H., A. D. Ba, D. Compagnon, L. B. Andonova, M. Paterson, M. Betsill, P. Newell, S. D. VanDeveer and T. Hale (2015), *Transnational Climate Change Governance*, Cambridge: Cambridge University Press.

Bulkeley, H., G. H. Edwards and S. Fuller (2014), 'Contesting Climate Justice in The City: Examining Politics and Practice in Urban Climate Change Experiments', *Global Environmental Change* 25, pp. 31–40.

C40 (2014), 'Climate Action in Megacities Volume 2.0', <http://c40.org/blog_posts/CAM2> (last accessed 9 August 2018).

C40 (2017a), 'The Cities', <http://www.c40.org/cities> (last accessed 9 August 2018).

C40 (2017b), 'C40', <http://c40.org/> (last accessed 9 August 2018).

CAIT (2017), 'CAIT Climate Data Explorer', <http://cait2.wri.org/historical> (last accessed 9 August 2018).

Caney, S. (2005), 'Cosmopolitan Justice, Responsibility, and Global Climate Change', *Leiden Journal of International Law* 18, pp. 747–75.

Caney, S. (2008), 'Global Distributive Justice and the State', *Political Studies* 56(3), pp. 487–518.

Caney, S. (2009), 'Climate Change and the Future: Discounting for Time, Wealth, and Risk', *Journal of Social Philosophy* 40(2), pp. 163–86.

Caney, S. (2010a), 'Climate Change and the Duties of the Advantaged', *Critical Review of International Social and Political Philosophy* 13(1), pp. 203–28.

Caney, S. (2010b), 'Climate Change, Human Rights, and Moral Thresholds', in S. M. Gardiner, S. Caney, D. Jamieson and H. Shue (eds), *Climate Ethics – Essential Readings*, Oxford: Oxford University Press, pp. 163–80.

Caney, S. (2011), 'Humanity, Associations, and Global Justice: In Defence of Humanity-Centred Cosmopolitan Egalitarianism', *The Monist* 94(4), pp. 506–34.

Caney, S. (2014), 'Two Kinds of Climate Justice: Avoiding Harm and Sharing Burdens', *The Journal of Political Philosophy* 22(2), pp. 125–49.

Caney, S., and Hepburn, C. (2011), 'Carbon Trading: Unethical, Unjust and Ineffective?', *Royal Institute of Philosophy Supplement* 69, pp. 201–34.

Carbon Brief (2016), 'Analysis: What Global Emissions In 2016 Mean for Climate Change Goals', <https://www.carbonbrief.org/what-global-co2-emissions-2016-mean-climate-change> (last accessed 9 August 2018).

Carbon Brief (2017), 'Analysis: Global CO_2 emissions set to rise 2% in 2017 after three-year "plateau"', <https://www.carbonbrief.org/analysis-global-co2-emissions-set-to-rise-2-percent-in-2017-following-three-year-plateau> (last accessed 9 August 2018).

Carbon Sequestration Leadership Forum (2017a), 'About Us', <https://www.cslforum.org/cslf/About-CSLF> (last accessed 9 August 2018).

Carbon Sequestration Leadership Forum (2017b), 'Charter', <https://www.cslforum.org/cslf/sites/default/files/documents/CSLFCharter2011.pdf> (last accessed 9 August 2018).

Carbon Trade Watch (2017), 'Climate Justice', <http://www.carbontradewatch.org/issues/climate-justice.html> (last accessed 9 August 2018).

Climate Action Tracker (2016), 'Effect of Current Pledges and Policies on Global Temperature', <http://climateactiontracker.org/global.html> (last accessed 9 August 2018).

Climate Funds Update (2017), 'The Data', <https://climatefundsupdate.org/data-dashboard/> (last accessed 9 August 2018).

Climate Group (2012), 'Lighting the Green Revolution' <http://www.theclimategroup.org/_assets/files/LED_report_web1%283%29.pdf> (last accessed 9 August 2018).

Climate Group (2017a), 'Who We Are', <http://www.theclimategroup.org/who-we-are/about-us/> (last accessed 9 August 2018).

Climate Group (2017b), 'Our Achievements', <http://www.theclimategroup.org/our-achievements/> (last accessed 9 August 2018).

Cochrane, A. (2013), 'Cosmozoopolis: The Case Against Group-Differentiated Animal Rights', *Law, Ethics and Philosophy* 1, pp. 127–41.

Collier, P. (2010), *The Plundered Planet: How to Reconcile Prosperity with Nature*, London: Allen Lane.

Costello, A., M. Abbas, A. Allen, S. Ball, R. Bellamy, S. Friel, N. Groce, A. Johnson, M. Kett, M. Lee, C. Levy, M. Maslin, D. McCoy, B. McGuire, H. Montgomery, D. Napier, C. Pagel, J. Patel, J. A. Puppim de Oliveira, N. Redclift, H. Rees, D. Rogger, J. Scott, J. Stephenson, J. Twigg, J. Wolff and C. Patterson (2009), 'Managing the Health Effects of Climate Change', *The Lancet* 373(9676), pp. 1693–733.

Dietzel, A. (2017), 'Global Justice' in S. McGlinchey, R. Walters and C. Scheinpflug (eds), *International Relations Theory*, Bristol: E-International Relations Publishing, pp. 91–6.

Earth Negotiations Bulletin (1997), 'Summary of the Third Conference of the Parties to the Framework Convention on Climate Change', <http://enb.iisd.org/download/pdf/enb1276e.pdf> (last accessed 9 August 2018).

Earth Negotiations Bulletin (2013), 'Summary of the Warsaw Climate Change Conference – November 2013', <http://enb.iisd.org/download/pdf/enb12594e.pdf> (last accessed 9 August 2018).

Earth Negotiations Bulletin (2014), 'Summary of the Lima Climate Change Conference – December 2014', <http://enb.iisd.org/download/pdf/enb12619e.pdf> (last accessed 9 August 2018).

Earth Negotiation Bulletin (2015a), 'Summary of the Geneva Climate Change Conference – February 2015', <http://enb.iisd.org/download/pdf/enb12626e.pdf> (last accessed 9 August 2018).

Earth Negotiations Bulletin (2015b), 'Summary of The Bonn Climate Change Conference', <http://enb.iisd.org/climate/unfccc/adp2-11/> (last accessed 9 August 2018).

Earth Negotiations Bulletin (2015c), 'Summary of the Paris Climate Change Conference', <http://enb.iisd.org/climate/cop21/enb/> (last accessed 9 August 2018).

Earth Negotiations Bulletin (2016), 'Summary of The Marrakech Climate Change Conference', <http://enb.iisd.org/download/pdf/enb12689e.pdf> (last accessed 9 August 2018).

Earth Negotiations Bulletin (2017), 'Summary of The Fiji/Bonn Climate Change Conference', <http://enb.iisd.org/download/pdf/enb12714e.pdf> (last accessed 9 August 2018).

Eckersley, R. (2012), 'Moving Forward in the Climate Negotiations: Multilateralism or Minilateralism?', *Global Environmental Politics* 12(2), pp. 24–42.

Estlund, D. (2014), 'Utopophobia', *Philosophy and Public Affairs* 42(2), pp. 113–34.

G7 (2015), 'A New Climate for Peace: Taking Action on Climate Change and Fragility Risks', <https://www.newclimateforpeace.org/> (last accessed 9 August 2018).

Gardiner, S. M. (2004a), 'Ethics and Global Climate Change', *Ethics* 114(3), pp. 555–600.

Gardiner, S. (2004b), 'The Global Warming Tragedy and the Dangerous Illusion of the Kyoto Protocol', *Ethics and International Affairs* 18(1), pp. 23–39.

Gardiner, S. M. (2010), 'Ethics and Climate Change: An Introduction', *Ethics and Climate Change* 1(1), pp. 54–66.

Goldenberg, S. (2014), 'G20: Obama to pledge up to $3bn to help poor countries on climate change', *The Guardian*, 14 November, <http://www.theguardian.com/environment/2014/nov/14/barack-obama-to-pledge-at-least-25bn-to-help-poor-countries-fight-climate-change> (last accessed 9 August 2018).

Hale, T. (2016), '"All Hands on Deck": The Paris Agreement and Nonstate Climate Action', *Global Environmental Politics* 16(3), pp. 12–22.

Hale, T., D. Held and K. Young (2013), *Gridlock: Why Global Cooperation is Failing When We Need It Most*, Cambridge, Durham: Polity Press.

Hansen, J., P. Kharecha, M. Sato, V. Masson-Delmotte, F. Ackerman, D. J. Beerling, P. J. Hearty, O. Hoegh-Guldberg, S.-L. Hsu, C. Parmesan, J. Rockstrom, E. J. Rohling, J. Sachs, P. Smith, K. Steffen, L. Van Susteren, K. Von Schuckmann and J. C. Zachos (2013), 'Assessing Dangerous Climate Change: Required Reduction of Carbon Emissions to Protect Young People, Future Generations and Nature', *Plos One* 8 (12), pp. 1–26.

Harris, P. (2010), *World Ethics and Climate Change: From International to Global Justice*, Edinburgh: Edinburgh University Press.

Harris, P. (2013), *What's Wrong with Climate Politics and How to Fix It*, Cambridge, Durham: Polity Press.

Harris, P., and T. Lee (2017), 'Compliance with Climate Change Agreements: The Constraints of Consumption', *International Environmental Agreements: Politics, Law, and Economics* 17(6), pp. 779–94.

Hayden, P. (2001), *The Philosophy of Human Rights*, St. Paul: Paragon House.

Hayden, P. (2010), 'The Environment, Global Justice and World Environmental Citizenship', in G. W. Brown and D. Held (eds), *The Cosmopolitanism Reader*, Cambridge, Durham: Polity Press, pp. 351–72.

Hayden, P. (2012), 'The Human Right to Health and the Struggle for Recognition', *Review of International Studies* 38(3), pp. 569–88.

Hayward, T. (2007), 'Human Rights versus Emission Rights: Climate Justice and the Equitable Distribution of Ecological Space', *Ethics and International Affairs* 21(4), pp. 431–50.

Heyward, C., and D. Roser, eds (2016), *Climate Justice in a Non-Ideal World*, Oxford: Oxford University Press.

Hof, A., M. Den Elzen and D. Van Vuuren (2010), 'Environmental Effectiveness and Economic Consequences of Fragmented Versus Universal Regimes: What Can We Learn From Model Studies?' in F. Biermann, P. Pattberg and F. Zelli (eds), *Global Climate Governance Beyond 2012: Architecture, Agency, and Adaptation*, Cambridge: Cambridge University Press, pp. 35–59.

Hoffman, M. J. (2011), *Climate Governance at the Crossroads: Experimenting with a Global Response after Kyoto*, Oxford: Oxford University Press.

Hoornweg, D., L. Sugar and C. L. T. Gómez (2011), 'Cities and Greenhouse Gas Emissions: Moving Forward', *Environment and Urbanization* 23(1), pp. 207–27.

Horta, O. (2013), 'Expanding Global Justice: The Case for the International Protection of Animals', *Global Policy* 4(4), pp. 371–80.

Hsu, A., A. S. Moffat, A. J. Weinfurter and J. D. Schwartz (2015), 'Commentary: Towards a New Climate Diplomacy', *Nature* 5, pp. 501–3.

Intergovernmental Panel on Climate Change (1990), 'First Assessment Report', <https://www.ipcc.ch/ipccreports/1992%20IPCC%20Supplement/IPCC_1990_and_1992_Assessments/English/ipcc_90_92_assessments_far_overview.pdf> (last accessed 9 August 2018).

Intergovernmental Panel on Climate Change (2001), 'Third Assessment Report: Summary for Policymakers', <https://www.ipcc.ch/pdf/climate-changes-2001/synthesis-syr/english/summary-policymakers.pdf> (last accessed 9 August 2018).

Intergovernmental Panel on Climate Change (2007), 'IPCC Fourth Assessment Report: Summary for Policy Makers', <https://www.ipcc.ch/pdf/assessment-report/ar4/syr/ar4_syr_spm.pdf> (last accessed 9 August 2018).

Intergovernmental Panel on Climate Change (2014a), 'IPCC Fifth Assessment Report: Summary for Policymakers', <https://www.ipcc.ch/pdf/assessment-report/ar5/wg3/ipcc_wg3_ar5_summary-for-policymakers.pdf> (last accessed 9 August 2018).

Intergovernmental Panel on Climate Change (2014b), 'Climate Change 2014: Mitigation of Climate Change' <http://www.ipcc.ch/pdf/assessment-report/ar5/wg3/ipcc_wg3_ar5_full.pdf> (last accessed 9 August 2018).

International Council on Human Rights Policy (2008), 'Climate Change and Human Rights: A Rough Guide', <https://www.ohchr.org/Documents/Issues/ClimateChange/Submissions/136_report.pdf> (last accessed 16 August 2017).

Jamison, D. T., J. G. Breman, A. R. Measham, M. Claeson, D. B. Evans, P. Jha, A. Mills and P. Musgrove (2006), *Disease Control Priorities in Developing Countries*, Oxford: Oxford University Press.

Kang, K. (2006), 'Climate Change and Human Rights', *Address to the Conference of the Parties of the UNFCCC*, 3–14 December.

Keohane, R. O., and D. G. Victor (2011), 'The Regime Complex for Climate Change', *Perspectives on Politics* 9(1), pp. 7–23.

Klein, R. J. T., and A. Persson (2008), 'Financing Adaptation to Climate Change: Issues and Priorities', *European Climate Platform Report* 8, pp. 1–13.

Kok-Chor, T. (2004), *Justice Without Borders*, Cambridge: Cambridge University Press.

Kotchen, M. (2017), 'Trump will stop paying into the Green Climate Fund. He has no idea what it is', *Washington Post*, 2 June, <https://www.washingtonpost.com/posteverything/wp/2017/06/02/trump-will-stop-paying-into-the-green-climate-fund-he-has-no-idea-what-it-is/?utm_term=.467acf7b4e7d> (last accessed 9 August 2018).

Labonte, R., and A. Ruckert (2014), 'The Social Determinants of Health', in G. W. Brown, G. Yamey and S. Wamala (eds), *Global Health Policy*, Cambridge: Wiley-Blackwell, pp. 267–85.

Lawford-Smith, H. (2016), 'Difference-Making and Individuals' Climate-Related Obligations', in Jeremy Moss (ed.), *Climate Change and Justice*, Cambridge: Cambridge University Press, pp. 64–83.

Lawrence, P. (2014), *Justice for Future Generations: Climate Change and International Law*, Cheltenham: Edward Elgar Publishing Ltd.

Limon, M. (2009), 'Human Rights and Climate Change: Constructing a Case for Political Action', *Harvard Environmental Law Review* 33, pp. 439–76.

Lomborg, B., (2001), *The Skeptical Environmentalist: Measuring the Real State of the World*, Cambridge: Cambridge University Press.

Maltais, A., and C. McKinnon (eds) (2015), *The Ethics of Climate Governance*, London: Rowman and Littlefield.

McKendry, C. (2016), 'Cities and the Challenge of Multiscalar Climate Justice: Climate Governance and Social Equity in Chicago, Birmingham, and Vancouver', *Local Environment* 21(11), pp. 1354–2371.

Möhner, A., and R. J. T. Klein (2007), 'The Global Environment Facility: Funding for Adaptation, or Adapting to Funds?', *Climate and Energy Programme, Working Paper*, Stockholm Environment Institute, pp. 1–18.

Moss, J., ed (2015), *Climate Change and Justice*, Cambridge: Cambridge University Press.

Miller, D. (2005), 'Reasonable Partiality Towards Compatriots', *Ethical Theory and Moral Practice* 8(1), pp. 63–81.

Miller, D. (2013), *Justice for Earthlings*, Cambridge: Cambridge University Press.

Müller, O., and M. Krawinkel (2002), 'Malnutrition and Health in Developing Countries', *Canadian Medical Association Journal* 173(3), pp. 279–86.

Myers, N. (2002), 'Environmental Refugees: A Growing Phenomenon of the 21st Century', *Philosophical Transactions of the Royal Society* 357(420), pp. 609–13.

Nussbaum, M. (2004), 'Beyond the Social Contract', *Oxford Development Studies* 32(1), pp. 3–18.

Nussbaum, M. (2006), *Frontiers of Justice: Disability, Nationality, Species Membership*, Cambridge, MA: Harvard University Press.

Okereke, C., and P. Coventry (2016), 'Climate Justice and the International Regime: Before, During, and After Paris', *WIREs Climate Change* 7(6), pp. 834–51.

Ostrom, E. (2009), 'A Polycentric Approach for Coping with Climate Change', *World Bank Policy Research Working Paper*, 5095, pp. 1–54.

Oxfam (2017), 'Extreme Carbon Inequality', <https://www.oxfam.org/en/research/extreme-carbon-inequality> (last accessed 9 August 2018).

Paddy, A. (2014), 'Countries Pledge $9.3bn for Green Climate Fund', *The Guardian*, 20 November, <http://www.theguardian.com/environment/2014/nov/20/countries-pledge-93bn-for-green-climate-fund> (last accessed 9 August 2018).

Page, E. (2012), 'Give it Up for Climate Change: A Defense of the Beneficiary Pays Principle', *International Theory* 4(2), pp. 300–30.

Pattberg, P. (2010a), 'Public–Private Partnerships in Global Climate Governance', *WIREs Climate Change* 1(2), pp. 279–28.

Pattberg, P. (2010b), 'The Role and Relevance of Networked Climate Governance', in F. Biermann, P. Pattberg and F. Zelli (eds), *Global Climate Governance Beyond 2012: Architecture, Agency, and Adaptation*, Cambridge: Cambridge University Press, pp. 146–64.

Pattinson, J. (2007), 'Humanitarian Intervention and International Law: The Moral Importance of an Intervener's Legal Status', *Critical Review of International Social and Political Philosophy* 10(3), pp. 301–19.

Pattinson, J. (2008), 'Whose Responsibility to Protect? The Duties of Humanitarian Intervention', *Journal of Military Ethics* 7(4), pp. 262–83.

Parfit, D. (2010), 'Energy Policy and the Further Future: The Identity Problem', in S. M. Gardiner, S. Caney, D. Jamieson and H. Shue (eds), *Climate Ethics – Essential Readings*, Oxford: Oxford University Press, pp. 112–21.

Pepper, A., (2017), 'Beyond Anthropocentrism: Cosmopolitanism and Nonhuman Animals', *Global Justice: Theory Practice Rhetoric* 9(2), pp. 114–33.

Pogge, T. (1989), *Realizing Rawls*, London: Cornell University Press.

Regional Greenhouse Gas Initiative (2017a), 'Welcome', <http://www.rggi.org/> (last accessed 9 August 2018).

Regional Greenhouse Gas Initiative (2017b), 'RGGI Benefits', <http://www.rggi.org/rggi_benefits> (last accessed 9 August 2018).

Risse, M. (2008), 'Who Should Shoulder the Burden? Global Climate Change and Common Ownership of the Earth', *Faculty Research Working Papers Series*, Harvard Kennedy School, RWP08-075, pp. 1–58.

Risse, M. (2012), *On Global Justice*, Princeton: Princeton University Press.

Sandberg, K. (2011), 'My Emissions Make No Difference: Climate Change and the Argument from Inconsequentialism', in *Environmental Ethics* 33(3), pp. 229–48.

Sander, C., H. Van Asselt, T. Hale, K. Abbot, M. Beisheim, M. Hoffman, B. Guy, N. Hoehne, A. Hsu, P. Pattberg, P. Pauw, C. Ramstein and O. Widerberg (2015), 'Reinvigorating International Climate Policy: A Comprehensive Framework for Effective Nonstate Action', *Global Policy* 6(4), pp. 466–73.

Sangiovanni, A. (2007), 'Global Justice, Reciprocity, and the State', *Philosophy and Public Affairs* 35(1), pp. 3–39.

Santos, M. (2017), 'Global Justice and Environmental Governance: An Analysis of the Paris Agreement', *Revista Brasileira de Política Internacional* 60(1), pp. 1–18.

Seyfang, G. (2005), 'Shopping for Sustainability: Can Sustainable Consumption Promote Ecological Citizenship?', *Environmental Politics* 14(2), pp. 290–306.

Shue, H. (1999), 'Global Environmental and International Inequality', *International Affairs* 75(3), pp. 533–7.

Shue, H. (2014), *Climate Justice: Vulnerability and Protection*, Oxford: Oxford University Press.

Singer, P. (1972), 'Famine, Affluence, and Morality', *Philosophy and Public Affairs* 1(1), pp. 229–43.

Singer, P. (2006), 'Ethics and Climate Change: A Commentary on MacCracken, Toman and Gardiner', *Environmental Values* 15(3), pp. 415–22.

Sinnott-Armstrong, W. (2010), 'It's Not My Fault: Global Warming and Individual Moral Obligations', in S. M. Gardiner, S. Caney, D. Jamieson and H. Shue (eds), *Climate Ethics – Essential Readings*, Oxford: Oxford University Press, pp. 332–46.

Stevenson, H., and J. S. Dryzek (2014), *Democratizing Global Climate Governance*, Cambridge: Cambridge University Press.

Stripple, J., and P. Pattberg (2010), 'Agency in Global Climate Change Governance: Setting the Stage', in F. Biermann, P. Pattberg and F. Zelli (eds), *Global Climate Governance Beyond 2012: Architecture, Agency, and Adaptation*, Cambridge: Cambridge University Press, pp. 137–45.

Tabuchi, H., and H. Fountain (2017), 'Bucking Trump, These Cities, States and Companies Commit to Paris Accord', *New York Times*, 1 June, <https://www.nytimes.com/2017/06/01/climate/american-cities-climate-standards.html?_r=0> (last accessed 9 August 2018).

Transition Towns (2017), 'What is Transition?', <http://www.transitiontowntotnes.org/about/what-is-transition/> (last accessed 9 August 2018).

UNEP (2015), 'Climate Commitments of Subnational Actors and Business: A Quantitative Assessment of Their Emission Reduction Impact', <http://apps.unep.org/redirect.php?file=/publications/pmtdocuments/-Climate_Commitments_of_Subnational_Actors_and_Business-2015CCSA_2015.pdf.pdf> (last accessed 9 August 2018).

UNEP (2017), 'Climate Change and Security Risks', <https://www.unenviron-ment.org/explore-topics/disasters-conflicts/what-we-do/risk-reduction/climate-change-and-security-risks> (last accessed 9 August 2018).

UNFCCC (1992), 'United Nations Framework Convention on Climate Change', <https://unfccc.int/resource/docs/convkp/conveng.pdf> (last accessed 9 August 2018).

UNFCCC (1997), 'Kyoto Protocol to The United Nations Framework Convention On Climate Change', <http://unfccc.int/kyoto_protocol/items/2830.php> (last accessed 9 August 2018).

UNFCCC (2015), 'Adoption of the Paris Agreement', <https://unfccc.int/documentation/documents/advanced_search/items/6911.php?priref=600008831> (last accessed 9 August 2018).

UNHCR (2017), 'Climate Change and Disasters', <http://www.unhcr.org/uk/climate-change-and-disasters.html> (last accessed 9 August 2018).

Vanderheiden, S. (2008), *Atmospheric Justice – A Political Theory of Climate Change*, Oxford: Oxford University Press.

Vanderheiden, S. (2015), 'Justice and Climate Finance: Differentiating Responsibility in the Green Climate Fund', *International Spectator* 50(1), pp. 31–45.

Verified Carbon Standard (2017a), 'Who We Are', <http://verra.org/about-verra/who-we-are/> (last accessed 9 August 2018).

Verified Carbon Standard (2017b), 'What We Do', <http://verra.org/about-verra/what-we-do/> (last accessed 9 August 2018).

Verified Carbon Standard (2017c), 'The VCS Program', <http://verra.org/project/vcs-program/> (last accessed 9 August 2018).

Vermeulen, S., B. M. Campbell and J. Ingram (2012), 'Climate Change and Food Systems', *Annual Review of Environment and Resources* 37, pp. 195–222.

Weiss, T. G. (2013), *Global Governance: Why? What? Whither?* Cambridge and Durham: Polity Press.

Whitman, J. (2009), *The Fundamentals of Global Governance*, Basingstoke: Palgrave Macmillan.

WHO (1946), 'Constitution of The World Health Organization', <http://www.who.int/governance/eb/who_constitution_en.pdf> (last accessed 9 August 2018).

WHO (2003), 'Climate Change and Human Health – Risks and Responses', <http://www.who.int/globalchange/publications/cchhbook/en/> (last accessed 9 August 2018).

WHO (2013), 'Climate Change and Health: A Tool to Estimate Health and Adaptation Costs', <http://www.euro.who.int/__data/assets/pdf_file/0018/190404/WHO_Content_Climate_change_health_DruckIII.pdf?ua=1> (last accessed 9 August 2018).

WHO (2017), 'Climate Change and Health', <http://www.who.int/mediacentre/factsheets/fs266/en/> (last accessed 9 August 2018).

Woodward, J. (1986), 'The Non-Identity Problem', *Ethics* 96(4), pp. 804–31.

World Bank (2010), 'World Development Report 2010 – Development and Climate Change', <http://www.worldbank.org/wdr2010> (last accessed 9 August 2018).

World Bank (2016), 'Gross Domestic Product 2016', <http://databank.world-bank.org/data/download/GDP.pdf> (last accessed 9 August 2018).

World Bank (2017), 'Cities and Climate Change: An Urgent Agenda', <http://sit-eresources.worldbank.org/INTUWM/Resources/340232-1205330656272/CitiesandClimateChange.pdf> (last accessed 9 August 2018).

INDEX

Abbott, Tony, 132, 163
Ability to Pay Principle (ATP),
    73, 74–9, 139–42, 145,
    158, 208
Ad Hoc Working Group on the
    Durban Platform for Enhanced
    Action (ADP), 144–8, 216
adaptation, 9–10
    Demand One, 88, 120–1, 128–38,
    175–82, 209
    and health, 54–5, 59–60, 65–6
    and less developed countries,
    211–12, 217
Adaptation Committee, 103, 135, 137,
    156, 158, 207–8
Adaptation Framework, 148
Adaptation Fund, 103, 131, 217
Addis Ababa, Ethiopia, 180
Africa, 7, 47, 176
African Charter on Human and
    People's Rights 1981, 51
Albania, 82
American Declaration of the
    Rights and Duties of Man
    1948, 51
Amoco, 82, 170
Angola, 69
animal rights, 57n
Annex I countries, 69, 142–3, 144, 151,
    157, 207, 216

Annex II countries, 69, 142, 144, 151,
    157, 207, 216
Annex X, 146
Antarctic, 8
Armenia, 82
Asia, 7, 47, 188
Asian Cities Climate Change
    Resilience Network (ACCCRN),
    104–5, 177–9, 181–2, 193, 194,
    198, 209
Asia-Pacific Partnership on Clean
    Development and Climate
    (APP), 105–6, 183–7, 193, 194,
    198, 209, 210
Australia, 82, 106, 132, 145, 170, 185,
    187, 200, 211

Ban Ki-Moon, 152–3
Bangladesh, 104, 178
Barry, Brian, 70–1
Beneficiary Pays Principle (BPP),
    76–7, 78
benefits and burdens, 14, 34–6, 79–80,
    105–6
Birmingham, 17
Brazil, 69, 72, 73, 143, 145,
    147–8, 184
BRICS countries, 68–9, 72, 73,
    184, 188
British Petroleum (BP), 82, 170

Bulkeley et al.
  and anti-consumerism, 195–6
  and APP, 185–6
  and city-based responses to climate
    change, 17
  and GHG emissions, 166
  and less developed countries,
    140, 187–8
  and mitigation and adaptation,
    175
  and non-state actors, 189
  and transnational actors, 187
  and transnational climate change
    governance, 162–3, 183, 199
Bush, George W., 72

C40, 82, 96–7, 103–4, 165, 169–74,
    193–4, 198, 209
California, 172
Canada
  APP, 106, 185
  COP20, 145
  CSLF, 184
  GHG emissions, 82, 170
  Kyoto Protocol, 124, 143, 147, 151
  transnational climate change
    governance, 187, 200, 211
Cancun Adaptation Framework, 135
Caney, Simon, 28, 31, 34–6, 39, 44,
    77–8, 87, 99–101, 113
Carbon Sequestration Leadership
    Forum (CSLF), 183–5, 187, 193,
    194, 198, 209, 210
Carbon Trade Watch, 196, 209, 211
carbon trading, 191–2
Charter of Fundamental Rights of the
    European Union 2000, 51
Chicago, 17
children, 55
China
  APP, 106, 185
  COP20, 145
  CSLF, 184
  GHG emissions, 32, 67, 150, 186
  Kyoto Protocol, 147–8

  less developed countries, 72, 82
  non-Annex I countries, 69, 143
  PPP, 73–4
cities, 82, 86, 96–7, 169–70, 171–2,
    179–81, 189, 193
Climate Action Network (CAN),
    168
climate change action, capability to
    enable, 99–101
Climate Change and Justice, 16
Climate Group
  GHG emissions, 198
  green technology, 107, 193
  hybrid transnational climate change
    governance, 97
  mitigation, 165
  state and non-state response, 194
  transnational climate change
    governance, 169, 171, 172–4, 209
*Climate Justice in a Non-Ideal
    World*, 16
Climate Justice Now, 168
Climate Convention 1992, 118–31,
    133–4, 138–59, 206
  Article 1.1, 122
  Article 3, 141–2
  Article 3.1, 121–2, 125, 140,
    149–50
  Article 3.14, 131, 133, 139
  Article 4.1e, 128–9
  Article 4.1f, 122
  Article 4.3, 103
  Article 4.4, 103, 129
  Article 4.5, 129
  Article 4.8, 130
  Article 11, 129
  Article 11.4, 129
  Article 15, 127
Climate Summit, 152
Cochrane, Alasdair, 57n
collective principles, 83
collective response, 10
Common but Differentiated
    Responsibility (CBDR), 105,
    138–42, 144–8, 157, 207

Conference of the Parties (COP), 95,
   97, 118
   adaptation, 103, 135
   as cross-fertilisation space, 214
   COP7, 134
   COP15, 134–135
   COP16, 135
   COP19, 144–145, 147
   COP20, 145, 147, 152–153
   COP21, 144–147, 153–155,
      157, 159, 168–169, 189,
      207–209, 216
   COP22, 160, 216
   COP23, 160, 215, 216–217, 217
   COP24, 217
   finance, 129
   and less developed countries, 140
   PATP model, 105
   transnational climate change
      governance, 168, 213
Connecticut, 192
constructive ambiguity, 121–2
consumers, 84–5
Copenhagen Accord 2009, 123,
   134–5
corporations, 82, 86–7, 172–3
cosmopolitan climate justice
   relational and non-relational
      divide, 26–8
   research, 15–17
cosmopolitan global justice, 1–3
cosmopolitanism, 12–17, 162–3
Coventry, Philip, 16

Dakar, Senegal, 180
Delaware, 192
Dell, 172
Demand One, 64–6, 88, 119, 164
   INDCs, 158
   multilateral climate change
      governance, 102–5, 120–38,
      155–6, 206–7
   non-relational moral minimum, 204
   transnational climate change
      governance, 165–2, 197–201, 209

Demand Two, 64–5, 88, 119, 164
   multilateral climate change
      governance, 105–6, 138–48, 155–8
   transnational climate change
      governance, 182–9, 197–8, 200–1,
      207, 210
Demand Three, 88, 120, 164
   multilateral climate change
      governance, 106–7, 148–55,
      155–7, 159
   transnational climate change
      governance, 189–96, 197–8,
      200–1, 210
diet, 85
disease, 48, 54, 62, 65, 121
displacement, 47, 49–50
Dominican Republic, 145
*Draft Declaration of Principles on
   Human Rights and Environment*, 44
droughts, 8

education, 55
Edwards, Gareth, 17
egalitarian non-relational theorists,
   27–28, 34–6
Estlund, David, 102, 115
The Ethics of Climate Governance, 16
Europe, 188
European Commission (EC), 184
European Union (EU), 13, 32, 140, 143
extreme weather events, 47
Exxon, 82, 170

fairness, 69–72
FedEx, 86
Figueres, Christina, 153
Fiji, 145
floods, 8, 48, 179
food security, 46, 62, 84–5
forced migration, 48
four-point hierarchy, 20, 114–15, 120,
   133, 155, 165, 197, 206
France, 82, 170
Fuller, Sara, 17
future generations, 14–15, 29–31, 60–4

G7, 46–7
G77, 153
Gardiner, Stephen, 15, 157, 159
GHG emissions
  APP, 185–6
  ATP, 75
  C40, 171–2
  carbon trading, 191–3
  cities, 169–70
  Climate Group, 173–4
  CSLF, 184
  definition, 22n
  Demand Three, 198
  fairness, 14
  global justice, 13
  and individuals, 81–2
  IPCC, 6, 9, 11, 64–5, 123–4
  Kyoto Protocol, 143
  and less developed countries, 67
  LMDCs, 146
  mitigation and adaptation, 102–4,
    166–8, 178, 211–12
  non-identity problem, 61–2
  PPP, 73–4
  UNFCCC, 95, 133
global collective action, 10
'global consumer class', 82
Global Investor Statement on
    Climate Change 2014, 176,
    181, 198, 209
global temperature rise
  adaptation, 88
  Demand One, 102–3, 120–1, 204
  future generations, 29
  human right to health, 53–4, 59, 64
  IPCC, 165
  Kyoto Protocol, 123–4
  mitigation, 145
  Paris Agreement, 125–6, 158, 208
Green Climate Fund (GCF), 132
green energy, 86–7, 101
green technology, 67, 74, 107
Greenland, 8
Greenpeace, 170

Haiti, 82
Hale, Thomas, 17, 98, 214
Hansen, James, 50
Harris, Paul, 15–16, 86, 155–6, 157,
    159, 218
Hayden, Patrick, 28, 44, 51
Hayward, Tim, 44–5
health, human right to, 42, 45–50,
    53–6, 59–66, 100, 158, 204, 214
heat stress, 48
heatwaves, 8
Heyward, Clare, 16
'High Level Action Day', 153
Hoffman, M. J., 98, 163, 175, 177,
    185, 187
Hoornweg, D., 82
house design, 65
Hsu, A., 166
Human Development Index, 68
Human Development Report 2008, 43
human right to health, 42, 45–50,
    50–3, 53–6, 59–66, 100, 158,
    204, 214
human rights, 28, 38, 42–5
'humanising climate change', 42–3
humanitarian intervention, 116n

IBM, 86
IKEA, 87, 172
India
  ACCCRN, 104, 178–9
  APP, 106, 185
  COP20, 145
  CSLF, 184
  GHG emissions, 67, 150
  green energy, 101
  less developed countries, 82
  non-Annex I countries, 69, 143
  PPP, 74
individuals, 81–2, 84–5
Indonesia, 104, 178
Indoor Temperature Comfort initiative,
    India, 179
'industrialised' countries, 69

injury, 48
Intended Nationally Determined
    Contributions (INDCs)
  adaptation, 135–7, 156
  APA, 216–17
  CBDR, 147–8
  Demand Three, 151–2, 154
  Demands One, Two and Three,
    157–9, 207–8
  Paris Agreement, 126–8, 145
  and USA, 168
Intergovernmental Panel on Climate
    Change (IPCC), 6, 7–8, 9–10,
    11–12, 53, 64–5, 120, 125, 165
International Council on Human
    Rights (ICHR), 43
International Covenant on Economic,
    Social and Cultural Rights, 51

Japan, 106, 124, 143, 145, 151, 184, 185
Justice for Future Generations, 16

Kang, Kyung-wha, 43
Kerry, John, 147
Kigali, Rwanda, 180
Kok-Chor, Tan, 37
Kolkata, 172
Kyoto Protocol
  APP, 105–6, 185
  Article 2.2, 142
  Article 10, 150–1
  Article 11, 142
  C40, 171
  CBDR, 138, 141–2, 147
  Climate Group, 172
  COPs, 95
  CSLF, 184
  Demand One, 123–7, 130–1, 156,
    157–8, 207–8
  Demand Three, 150–1, 159
  GCF, 132
  GHG emissions, 102–4
  multilateral climate change
    governance, 2–3, 97, 118–20

non-Annex I countries, 69
  Singer, Peter, 100
  technology transfer, 133–4
  transnational climate change
    governance, 183
  USA, 67, 72, 193

Lawrence, Peter, 16, 157, 159
Least Developed Countries, 69,
    72–5, 77–80, 134, 140, 148,
    178, 180
Least Developed Countries Fund, 131
Least Developed Country Expert
    Group (LEG), 134, 137, 156, 158,
    207–8
Least Developed Country Work
    Programme, 148, 158–9, 208
LED Lighting Project, 86, 107, 173
less developed countries
  adaptation, 59–60, 66, 175–6
  benefits and burdens, 6–7
  categorisation, 68–9, 143
  CSLF, 184
  Demand Two, 200
  fairness, 67, 71–20
  inclusion in climate justice, 31–7
  Kyoto Protocol, 131–3
  mitigation, 145
  Paris Agreement, 217
  PPP, 73–5, 76, 140
  'well-being', 51–2
'life worth living', 62–3
Like Minded Group of Developing
    Countries (LMDCs), 145–6
'Lima Action Plan', 145–6
'Lima Call for Action', 145
Lima-Paris Action Agenda (LPAA),
    153–4
living conditions, 55

McKendry, Corina, 17
McKinnon, Catriona, 16
Maine, 192
malnutrition, 48

Maltais, Aaron, 16
Mandalay, Myanmar, 180
Maryland, 192
Massachusetts, 192
Mexico, 184
Middle East and North Africa
    (MENA), 188, 200, 211
Miller, David, 33, 38
misinformation, 11
mitigation, 9–10, 29, 33, 54, 88, 121–8,
    145, 166–5, 217–18
Mobil, 82, 170
Moss, Jeremy, 16
multilateral (state), 2–3
Mumbai, 172
Myers, N., 47

National Adaptation Plans (NAPs),
    135, 137, 148, 156, 157, 207
National Adaptation Programmes of
    Action (NAPAs), 134–5, 136, 137,
    148, 156, 157, 158, 207–8
nationals and non-nationals, 38
Netherlands, 82, 170
*A New Climate for Peace: Taking
    Action on Climate and Fragility
    Risks*, 46–7
New Hampshire, 192
New York, 192
New York Climate Summit 2015,
    166, 176
New York Declaration on Forests, 176
New Zealand, 7, 124, 143, 145, 147,
    151, 184
Newell, Peter, 140
non-Annex I countries, 69, 143, 180
non-governmental organisations
    (NGOs), 177
non-identity problem, 61–2, 63
non-relational and relational, 28–40
non-relational moral minimum, 60
Nonstate Actor Zone for Climate
    Action (NAZCA), 153–4
non-state actors, responsibilities of,
    80–8, 192–3

North America, 7, 188
nourishment, 53, 54, 56, 62
Nussbaum, Martha, 52–3, 56

Obama, Barack, 132
Oceania, 188, 200, 211
Okereke, Chukwumerije, 16
Olympic Village Micro-Energy
    Building, 186
100 Resilient Cities (100RC), 177–8,
    179–82, 193, 194, 198, 209
Organisation for Economic
    Co-operation and Development
    (OECD), 69, 142, 169
Oxfam, 82

Parfit, Derek, 61–2
Paris Agreement
    adaptation, 103, 135–6
    ADP, 144–8
    Article 2, 136
    Article 4.4, 32–3, 146
    Article 7, 136
    Article 7.1, 136
    Article 7.14, 136
    Article 7.6, 136
    Article 7.7, 136
    Article 9, 136, 137
    Article 9.1, 136
    Article 9.9, 136–137
    Article 10, 136, 137
    Article 10.1, 137
    Article 10.4, 137
    CBDR, 138
    Demand One, 158
    Demand Three, 106, 155–6
    Demand Two, 157
    Demands One, Two and Three,
        206–9
    multilateral climate change
        governance, 97–8, 118–20, 125–8,
        159, 168, 216
    non-state actors, 150, 152, 154, 212
    PATP model, 142–4
    temperature, 95, 217

transnational climate change
governance, 2–3, 16–17, 163
USA, 67, 72, 100–1, 184, 193
Pattberg, P., 187
Pattinson, J., 116n
Paynesville, Liberia, 180
Petro China, 150
Philippines, 104, 178
Polluter Pays Principle (PPP), 73–4,
76–8, 139–42, 145–6, 148,
158, 208
Polluter's Ability to Pay (PATP) model,
58–9, 73
APP, 186–7
CBDR, 138–9, 141–2
CSLF, 184–5, 210
Demand Two, 88, 105–6, 182–3,
198, 201
INDCs, 147, 158
Kyoto Protocol, 208
less developed countries, 143–4
LMDCs, 146
states, 78–80
pragmatic argument, 63–4
private sector, 87
private transnational climate change
governance, 97
promotion of public health, 65
public projects, 183–4
public transnational climate change
governance, 96–7

Rawls, John, 28
refugees, 47–8
Regional Greenhouse Gas Initiative
(RGGI), 191, 192–3, 194, 198,
209, 210
Rhode Island, 192
Rio de Janeiro, 172
Rio+20 summit 2012, 217–18
Risse, Mathias, 31, 76–8
Rockefeller Foundation, 179
Roser, Dominic, 16
Russia, 124, 143, 151, 184
Rwanda, 69

safe water, 49
Samoa, 134
Sandberg, K., 83
Sander, C., 169
Santos, Marcelo, 16
Saudi Arabia, 82, 170
Scarborough, 85
scientific uncertainty, 11–12
sea levels, 8, 47
Shell, 82, 170
shelter, 53, 54, 56, 62
Singer, Peter, 85–6, 99–100
Sinnott-Armstrong, Walter, 81, 82–3
small island communities, 7
social factors, 55
Somalia, 69
South Africa, 69, 143, 184
South Australia, 172
South Korea, 106, 185
Spain, 82, 170
Special Climate Change Fund, 131
states, responsibilities of, 67–80
Sub-Saharan Africa (SSA), 188,
200, 211
sufficitarian non-relational theorists,
27–8, 31, 33
Switzerland, 145
Syrian refugee crisis, 47

Taobao, 172
Telemar, 150
Texaco, 82, 170
Thailand, 104, 145, 178, 179
'tipping point', 63–4
Toronto, 172
'total universe', 96, 162
Transition Towns, 196, 209, 211
Trump, Donald, 1, 67, 72, 127, 132, 168
Type II Partnerships, 217–18

Uganda, 69
United Nations Development
Programme (UNDP), 68
United Nations Environment
Programme (UNEP), 169, 176

United Nations Framework
    Convention on Climate Change
    (UNFCCC)
ambiguity, 113–14
BRICS countries, 188
categorisation, 69, 178, 180
COP21, 153–4
Demand One, 102–3, 120–8, 130–1,
    133–4, 198, 209
Demand Three, 106, 149–50,
    152, 201
Demand Two, 105–6, 138–9, 141, 144
Demands One, Two and Three, 160
GCF, 132
GHG emissions, 168
international authority, 108–12,
    115–16, 116n
Kyoto Protocol, 157
less developed countries, 32–3
mitigation, 147
multilateral climate change
    governance, 2, 16, 95–6, 118–20,
    205–6, 215–18
non-state actors, 214
Paris Agreement, 156, 159
states, 80, 101
transnational climate change
    governance, 190, 212
United Nations High Commissioner
    for Refugees (UNHCR), 47–8
United Nations Secretary General
    (UNSG), 152–3
United Nations (UN), 43, 140, 170
United States of America (USA)
    ADP, 145–6
    APP, 106, 185
    ATP, 75
    COPs, 140

CSLF, 184
GCF, 132
GHG emissions, 13, 82
INDCs, 147
Kyoto Protocol, 72, 124, 143,
    147–8, 151
Paris Agreement, 67, 72, 100–1,
    127, 168
RGGI, 192–3
transnational climate change
    governance, 11, 187, 200, 211
Universal Declaration of Human
    Rights (UDHR), 44–5, 50
    Article 1, 46
    Article 13, 47
    Article 25, 44, 49–50, 51–2

Vancouver, 17
Vanderheiden, Steven, 16, 28, 60,
    71–2, 140
Venezuela, 145, 153
Verified Carbon Standard (VCS), 97,
    191–2, 194, 198, 209, 210
Vermont, 192
Vietnam, 104, 178

warming, 7–8, 9, 54
water shortages, 46, 49, 65
water supplies, 62
*What's Wrong with Climate Politics
    and How to Fix It*, 15–16
World Bank, 31, 68–9, 143
World Ethics and Climate Change,
    15–16
World Health Organization (WHO),
    46, 49, 50–1
World Summit on Sustainable
    Development 2002, 217–18